DATE DUE

APR 0 9 1994	
JAN 2 0 2004	

Population and Community Biology

STAGE-STRUCTURED POPULATIONS

Population and Community Biology Series

Principal Editor

M. B. Usher
Reader, Department of Biology, University of York, UK

Editors

M. L. Rosenzweig
Professor, Department of Ecology and Evolutionary Biology, University of Arizona, USA

R. L. Kitching
Professor, Department of Ecosystem Management, University of New England, Australia

The study of both populations and communities is central to the science of ecology. This series of books will explore many facets of population biology and the processes that determine the structure and dynamics of communities. Although individual authors and editors are given freedom to develop their subjects in their own way, scientific rigour is essential and often a quantitative approach to analysing population and community phenomena will be used.

Already published
Population Dynamics of Infectious Diseases: Theory and applications
Edited by R. M. Anderson

Competition
Paul A. Keddy

Multivariate Analysis of Ecological Communities
P. Digby and R. Kempton

The Statistics of Natural Selection
Bryan F. J. Manly

Food Webs
Stuart L. Pimm

Predation
Robert J. Taylor

STAGE-STRUCTURED POPULATIONS

Sampling, analysis and simulation

Bryan F. J. Manly

University of Otago
New Zealand

London   New York
CHAPMAN AND HALL

First published in 1990 by
Chapman and Hall Ltd
11 New Fetter Lane, London EC4P 4EE
Published in the USA by
Chapman and Hall
29 West 35th Street, New York NY 10001

© *1990 Bryan F. J. Manly*

Typeset in 10/12 Times by
Keyset Composition, Colchester
Printed in Great Britain by
St. Edmundsbury Press Ltd
Bury St. Edmunds, Suffolk

ISBN 0 412 35060 2

British Library Cataloguing in Publication Data

Manly, Bryan F. J. 1944–
 Stage-structured populations.
 1. Organisms. Population. Dynamics. Statistical models
 I. Title   II. Series
 574.5'248

 ISBN 0-412-35060-2

Library of Congress Cataloging in Publication Data

Manly, Bryan F. J., 1944–
 Stage-structured populations: sampling, analysis, and simulation
 Bryan F. J. Manly.
  p.   cm. – (Population and community biology series)
 Includes bibliographical references.
 ISBN 0-412-35060-2
 1. Population biology – Statistical methods. 2. Population
 biology – Mathematical models. I. Title. II. Series.
 QH352.M35   1990
 574.5'248'072 – dc20   89-23878 CIP

. . . but time and chance happen to them all.

Ecclesiastes

To my mother

Contents

Preface   xi

1 Stage-structured populations   1
  1.1 Introduction   1
  1.2 Stage-frequency data   2
  1.3 Key factor analysis   5
  1.4 Case studies   6

2 Sampling for population estimation   7
  2.1 Introduction   7
  2.2 Populations and samples   7
  2.3 Simple random sampling   8
  2.4 Determining sample sizes with simple random sampling   12
  2.5 Stratified random sampling   13
  2.6 Ratio estimation   18
  2.7 Regression estimation   22
  2.8 Cluster sampling   23
  2.9 Systematic sampling   23
  2.10 Multi-stage sampling   24
  2.11 Sampling for stage-frequency data   24
  2.12 Sampling species assemblages   29
  2.13 Special sampling methods   30
       Exercises   32

3 Maximum likelihood estimation of models   33
  3.1 The method of maximum likelihood   33
  3.2 Models for count data   33
  3.3 Computer programs   36
  3.4 Measuring goodness-of-fit   37
  3.5 Comparing models   38
  3.6 The heterogeneity factor   39

4 Analysis of multi-cohort stage-frequency data   42
  4.1 Multi-cohort stage-frequency data   42
  4.2 Temperature effects   44

4.3   Effect of mortality on stage durations   45
4.4   Methods for analysing multi-cohort stage-frequency data   46
4.5   Assessing estimates by simulation   46
4.6   The Kiritani–Nakasuji–Manly (KNM) method of analysis   52
4.7   The KNM method with iterative calculations   63
4.8   The Kempton method of estimation   64
4.9   Variations of the Kempton type of model   73
4.10 The Bellows and Birley model   79
4.11 Comparison of models   83
      Exercises   83

5 Analysis of single cohort stage-frequency data   86
5.1   Types of single cohort data   86
5.2   Analysis using multi-cohort methods   86
5.3   Data without mortality   90
5.4   Non-parametric estimation   94
5.5   Parametric models for estimation   96
5.6   Estimating the durations of stages   97
      Exercises   99

6 Matrix and other models for reproducing populations   101
6.1   Continually reproducing populations   101
6.2   The Bernardelli–Leslie–Lewis model   101
6.3   Lefkovitch's model for populations grouped by life stages   103
6.4   Usher's model   105
6.5   Further generalizations and extensions   108
6.6   Sampling variation and other sources of errors   109
6.7   Discussion   114
6.8   Other models for populations with continuous recruitment   115
      Exercise   117

7 Key factor analysis   118
7.1   Populations observed over a series of generations   119
7.2   Density-dependent survival   121
7.3   The Varley and Gradwell (1960) graphical method of key
      factor analysis   122
7.4   Extensions to the Varley and Gradwell approach   124
7.5   Detecting density-dependent k values   128
7.6   The Manly (1977) model for key factor analysis   132
7.7   The relative merits of different methods of key factor
      analysis   138
7.8   Using simulation with key factor analysis   139
7.9   Testing for delayed density-dependent mortality   148

7.10 Recent developments	150
7.11 Computer program	151
Exercise	151
8 Case studies	**152**
8.1 Introduction	152
8.2 The sheep blowfly *Lucilia cuprina*	152
8.3 The nematode *Paratrichodorus minor*	159
8.4 The pink cotton bollworm moth *Pectinophora gossypiella*	162
8.5 The southern pine beetle *Dendroctonus frontalis*	163
8.6 The grey pup seal *Halichoerus grypus*	166
References	168
Author index	181
Subject index	184

ERRATA

p. 77, line 12 should read
… development variable. The description 'inverse normal' does not come about …

p. 80, line 5 should read
… distribution of the duration of a stage reflects the …

Stage-Structured Populations
Bryan F.J. Manly
Published by Chapman and Hall, 11 New Fetter Lane, London EC4P 4EE
ISBN 0 412 35060 2

Preface

This book provides a review of methods for obtaining and analysing data from stage-structured biological populations. The topics covered are sampling designs (Chapter 2), the estimation of parameters by maximum likelihood (Chapter 3), the analysis of sample counts of the numbers of individuals in different stages at different times (Chapters 4 and 5), the analysis of data using Leslie matrix types of model (Chapter 6) and key factor analysis (Chapter 7). There is also some discussion of the approaches to modelling and estimation that have been used in five studies of particular populations (Chapter 8).

There is a large literature on the modelling of biological populations, and a multitude of different approaches have been used in this area. The various approaches can be classified in different ways (Southwood, 1978, ch. 12), but for the purposes of this book it is convenient to think of the three categories mathematical, statistical and predictive modelling.

Mathematical modelling is concerned largely with developing models that capture the most important qualitative features of population dynamics. In this case, the models that are developed do not have to be compared with data from natural populations. As representations of idealized systems, they can be quite informative in showing the effects of changing parameters, indicating what factors are most important in promoting stability, and so on.

Statistical modelling is driven by the needs of data analysis. Often, practical considerations will dictate the type of data that can be collected on a population and the primary interest is in using the data to estimate important population parameters. The models used to this end are as simple or as complicated as is needed by the circumstances. A model may be known to be unrealistic in some respects but it can still be useful if this does not seriously affect the estimates of crucial parameters.

Predictive modelling is generally carried out to solve some real practical problem. For example, a population of insects may have the potential to cause a great deal of damage to the trees in a forest and it is important for forest managers to have some means of predicting the size of the insect population in order to decide what control measures, if any, should be taken. This type of model is often necessarily rather complicated, consisting of submodels for different components of the population, with the parameters of these submodels estimated from field data and laboratory experiments. It is used primarily to predict population sizes, but some of the parameters used in the model may be estimated by matching predictions with field data.

This book is concerned almost entirely with statistical modelling. Most of the models that are discussed are introduced solely because of their potential use for analysing data in order to gain information about specific populations. In some cases, the models have been used or could be used for theoretical investigations of general principles of population dynamics. However, these applications are mentioned only briefly with no attempt at full referencing. Some examples of predictive models are discussed, but only to emphasize the uses of the simpler statistical models as part of these more complicated ones.

Most of the calculations for my examples have been carried out using two packages of programs that have been written for the IBM PC and compatible computers. One covers the methods of analysis for stage-frequency data that are discussed in Chapters 4–6. The other is for key factor analysis using the methods discussed in Chapter 7. For more information about the cost and availability of these programs, readers should write to me at the address shown below.

Many people have helped me, directly and indirectly, in the writing of this book. Particularly, I wish to thank Patricia Munholland for commenting on Chapter 2, and all the participants of the Laramie conference on the Estimation and Analysis of Insect Populations held in January 1988. The discussions and papers presented at that conference have influenced every chapter that follows. Thanks are also due to the United States–New Zealand Cooperative Science Program for supporting my attendance at the conference, the University of Otago Research Committee for financial support for computing, and the Department of Statistics of the University of Wyoming for facilities to complete the manuscript while I was on sabbatical leave during 1988.

Bryan F. J. Manly

Department of Mathematics and Statistics
University of Otago, PO Box 56
Dunedin, New Zealand

February 1989

1 Stage-structured populations

1.1 INTRODUCTION

The study of the dynamics of natural biological populations is often hampered by the difficulty of determining the ages of individuals. However, this problem can sometimes be circumvented by working in terms of a stage structure rather than an age structure. The life cycle of the individuals in a population may consist of a series of recognizable morphological stages that are entered, one after another, until death. It is then possible to model the dynamics of such a population in terms of the distributions of the durations of stages and temporal survival rates.

Insect populations provide a major area of application of the stage-structured approach to modelling. Many insects pass from an egg stage into several instar stages and then into the adult stage, if they survive that long. Typically, there are between 4 and 20 immature stages, each of them having a mean duration of a few days. The adults usually live from a few days to a few weeks. Similar life cycles are found in crustaceans, nematodes, echinoderms and various other organisms.

In this book, attention will be restricted to cases where individuals cannot miss a stage. Therefore, with q stages, development is always in the order $1 \rightarrow 2 \rightarrow 3 \ldots q - 1 \rightarrow q$, providing that death does not intervene. Further, a definite maturation process will usually take place in stages, so that a certain minimum time is required to be spent in each one. Some models assume that each individual spends the same time in each stage. This is unlikely to be true. However, such a model may be a better approximation than one which assumes that individuals can leave a stage immediately after entering it. For most of the situations that will be considered in this book, death can occur in any stage. Losses from stages 1 to $q - 1$ will then be due to individuals developing to the next stage, migrating or dying. All losses from the final stage will be by migration or death. The survival rate per unit time may or may not vary from stage to stage and/or with time.

The remainder of this first chapter consists primarily of an overview of the topics included in the chapters that follow, with examples to illustrate the types of data that are available for analysis.

1.2 STAGE-FREQUENCY DATA

Often, with field populations the only information that can be collected easily will be counts or estimates of the numbers in different stages at different points in time. This is called **stage-frequency data**, and takes the form indicated in Table 1.1. Ideally, the sampling fraction should be constant for the different sampling times so that apart from sampling errors the data reflect the changing frequencies in the population.

There are various reasons why it might be important to collect and analyse stage-frequency data. In some cases, it is a question of a research interest in an organism. At other times, it might be a matter of comparing the population dynamics of several similar species in the same area. This was the case with Bradley's (1985) study of four grasshopper populations on a common site. Often, data are collected to aid in pest management. For example, such data are needed to provide parameter estimates for a phenology modelling system such as PETE (Welch *et al.*, 1978; Croft *et al.*, 1980), which is designed to monitor key events in the field so that farmers can be given appropriate recommendations about control measures.

Determining stage-frequencies for a natural population may simply involve counting the individuals in an area. However, it is quite likely that a rather more complicated sampling design will be required involving, for example, stratification of the area covered by the population. Much of the standard theory of sampling designs then becomes relevant. This theory is therefore reviewed in Chapter 2. Also in Chapter 2 is a brief discussion of some special sampling methods that are useful for estimating frequencies in cases where direct counts are impossible.

Many models for stage-frequency data provide equations for the expected values of the frequencies in terms of parameters for survival, stage durations and the rate of entry to stage 1. In principle, the parameters can then be estimated by the method of maximum likelihood, given appropriate

Table 1.1 Stage-frequency data where f_{ij} denotes the sample count or estimate of the number of individuals in stage j at time t_i

Sample time	Sample frequency in stage				
	1	*2*	*3*	. . .	*q*
t_1	f_{11}	f_{12}	f_{13}	. . .	f_{1q}
t_2	f_{21}	f_{22}	f_{23}	. . .	f_{2q}
.					
.					
.					
t_n	f_{n1}	f_{n2}	f_{n3}	. . .	f_{nq}

assumptions about the distribution of the observed frequencies about their expected values. This method of estimation is discussed in Chapter 3 for the case where these distributions are Poisson so that sampling variances are equal to expected values. The possibly more common case where variances are proportional to expected values is also considered because this can be handled just like the Poisson case, with the variance inflation (or reduction) being estimated by a heterogeneity factor.

Tables 1.2 and 1.3 show two examples of stage-frequency data. The first was obtained by sampling a population of the grasshopper *Chorthippus parallelus* while it was developing in summer and autumn on East Budleigh Common in Devonshire, UK (Qasrawi, 1966). Here, interest was in daily survival rates, numbers entering stages, and the time spent in stages. The second concerns stages of breast development of New Zealand schoolgirls (Tanner, 1962; McCullagh, 1983). In this case, deaths are negligible and the sample sizes at different times are arbitrary rather than being proportional to the total population size. This example is somewhat different from most of those that will be considered in the following pages but still fits within the framework being considered.

A large number of special methods of analysis have been developed for data such as those in Table 1.2, and these are considered in Chapter 4. Some of these methods can also be applied to the breast development data of Table 1.3, where mortality is not considered. However, with the breast development data some more standard statistical methods are also applicable, as discussed in Chapter 5. A further distinction between Chapters 4 and 5 is that their primary concerns are with multi-cohort and single cohort data, respectively. Multi-cohort data come from studying populations where individuals enter the first development stage over an extended period of time. With single cohort data the entries to this stage occur at more or less the same time. The difference between these situations is that with multi-cohort data it is often necessary to estimate a distribution of entry times to stage 1, which makes the estimation of other parameters more complicated.

There was no continuous recruitment with the populations from which the data in Tables 1.2 and 1.3 were drawn. However, in some cases samples are taken from reproducing populations where the individuals entering stage 1 are produced entirely by the individuals already in the population. An example is shown in Table 1.4. Here there are four stages (eggs, larvae, pupae and adults), with the adult females producing the eggs. The population was set up by Lefkovitch (1964a) with 80 adults in four separate cages of 20 adults. With this type of population there is interest in producing a model that can be used to predict the changes in stage-frequencies that will follow when the population starts with specified frequencies in stages. Some approaches based on using a transition matrix for the changes from one sample time to the next are discussed in Chapter 6. Some simpler analyses based on the ratio of eggs to females are also considered in that chapter.

Table 1.2 Sample counts of grasshoppers (*Chorthippus parallelus*) in four instar stages and the adult stage at East Budleigh Common, Devonshire in 1964 (from Qasrawi, 1966). The sampling fraction was 0.002 for 29 May–23 September. The sampling fraction was 0.00143 on 20 and 25 May so the counts on these two days have been multiplied by the ratio 0.002/0.00143 = 1.4 to give adjusted counts that are comparable with the other ones

Date	Day	Instar 1	Instar 2	Instar 3	Instar 4	Adult	Total
May 20	4	7.0*					7.0
25	9	8.4*					8.4
29	13	14					14
June 3	18	10	1				11
10	25	7	5	1			13
15	30	1	10				11
18	33	1	8	1	1		11
22	37	3	8	4	2		17
25	40	7	12	6			25
29	44		7	6	6		19
July 2	47	1	1	6	4	1	13
9	54		1	3	2	1	7
13	58		4	4	4	5	17
16	61			1	3	2	6
20	65		1	1	5	6	13
24	69		1	1	2	5	9
27	72					6	6
31	75					6	6
Aug. 5	81				1	6	7
11	87				1	1	2
14	90					3	3
19	95					3	3
24	100					5	5
28	104					3	3
Sept. 2	109					4	4
8	115					2	2
11	118					2	2
17	124					2	2
23	130					1	1

*Adjusted counts.

Table 1.3 Breast development in a survey of New Zealand schoolgirls. The stages represent five levels of development, in order

Age (years)	Developmental stage					Total
	1	2	3	4	5	
10–11	621	251	50	7	0	929
11–12	292	353	214	72	5	936
12–13	132	273	337	160	39	941
13–14	50	182	397	333	132	1094
14–15	27	69	273	501	289	1159

Table 1.4 Stage-frequencies for a population of the cigarette beetle *Lasioderma serricorne* set up as Experiment II by Lefkovitch (1964a), and allowed to develop for 105 weeks. These are totals from four replicates started with 20 adults each

Week	Eggs	Larvae	Pupae	Adults	Total
0	0	0	0	80	80
3	0	1 355	0	0	1 355
6	4 671	53	23	1 008	5 755
9	3	7 545	2	12	7 562
12	117	1 586	661	783	3 147
15	1 196	4 749	13	855	6 813
18	55	2 838	551	135	3 579
21	2 023	2 488	54	704	5 269
24	450	4 117	82	328	4 977
27	1 001	2 901	288	659	4 849
30	847	3 916	54	446	5 263
33	534	1 834	254	516	3 138
36	1 177	3 883	112	526	5 698
39	156	2 937	278	297	3 668
42	1 206	3 168	122	594	5 090
45	389	5 258	129	457	6 233
48	358	3 514	218	275	4 365

1.3 KEY FACTOR ANALYSIS

Another type of information is obtained by estimating the total numbers entering different stages for a series of successive generations of a population. An example is shown in Table 1.5 for the pine looper *Bupalus piniarius* at Hoge Veluwe, The Netherlands (Klomp, 1966). Here, the nature of some

Table 1.5 Numbers per square metre entering stages for a population of the pine looper *Bupalus piniarus*. The stages are: 1, potential eggs (reproducing females, ×216); 2, eggs in July; 3, larvae in August; 4, larvae in September; 5, nymphs in November; 6, pupae in December; 7, pupae in April of the following year; 8, moths in June; 9, female moths; and 10, reproducing females

					Stage					
Year	1	2	3	4	5	6	7	8	9	10
1953	7.7	6.3	4.5	4.5	2.5	3.0	3.0	1.1	0.52	0.30
1954	64	61	12	12	3.5	3.3	3.0	1.4	0.61	0.19
1955	41	33	14	12	3.2	2.7	2.6	1.5	0.67	0.35
1956	76	58	15	12	4.1	1.8	1.7	0.90	0.47	0.14
1957	29	22	1.8	0.80	0.73	0.13	0.12	0.10	0.04	0.025
1958	5.4	5.4	3.6	2.2	1.8	1.0	0.87	0.48	0.21	0.11
1959	25	24	5.8	3.7	1.9	1.2	1.1	0.58	0.28	0.15
1960	31	28	7.5	6.5	3.6	2.3	2.3	1.5	0.79	0.31
1961	67	61	26	26	6.2	3.5	3.4	2.1	0.86	0.59
1962	127	99	25	20	8.4	5.5	5.0	1.1	0.54	0.26
1963	56	47	18	18	8.6	4.9	4.6	0.83	0.44	0.24
1964	52	42	6.5	3.7	1.6					

of the 'stages' may be different from the developmental stages of stage-frequency data because they are based on populations rather than on individuals. It is, for example, perfectly satisfactory for stage 9 for the pine looper population to be the number of moths in the population that enter the stage of being female in June. Questions of interest in a case like this concern the extent to which the mortality rates in different stages are density-dependent, which stages contribute most to the overall variation in numbers, and what can be said about the stability of the population. This type of study, which is called key factor analysis, is discussed in Chapter 7.

1.4   CASE STUDIES

The methods covered in Chapters 2–7 are general in the sense that they are widely applicable and do not rely heavily on the characterisics of particular populations. An alternative to this type of analysis involves making a serious attempt to model the special features of the population of interest. The result will then be more complex but also potentially more realistic. In some cases simulation will be the only realistic way to proceed, and parameter estimation will involve a combination of field sampling, the use of laboratory experiments, and matching simulated results with real data. Some examples of modelling particular populations of insects, nematodes and grey seals are discussed in Chapter 8.

2 Sampling for population estimation

2.1 INTRODUCTION

Studies of natural populations of animals and plants almost inevitably involve sampling programmes. Sometimes these programmes are relatively straightforward. For example, to determine the proportion of trees that are infested in a large orchard of numbered trees, a random sample of tree numbers can be drawn and the chosen trees inspected. The sample proportion of infested trees is then an unbiased estimator of the orchard proportion and standard statistical methods allow limits to be put on the likely error of estimation. However, the situation is much more complicated if the problem is to estimate the number of insects in different life stages in a population that covers a large heterogeneous area. In this case, the different stages are likely to occupy different parts of the habitat and sampling all stages with the same intensity may be difficult.

To design and carry out good sample surveys for population estimation requires ability in three areas of expertise: (1) the standard theory of sampling for the estimation of means, proportions and totals (using simple random sampling, stratified sampling, cluster sampling, etc.); (2) special methods used for ecological studies (mark recapture sampling, line transect sampling, etc.); and (3) special devices used for sampling particular types of habitat (suction traps, sieves, light traps, etc.). This chapter provides a review of the first two of these areas. The third area is covered only incidentally in examples, and a more specialized source such as Southwood (1978) should be consulted for more information on this topic.

2.2 POPULATIONS AND SAMPLES

Before carrying out any sampling programme it is important to be clear about the population that is being studied. A distinction must often be made between the **target population** (the one that is really of interest), and the **sampled population**. Obviously, the target and sampled population should ideally be the same, but this is sometimes impossible to achieve. It can be considered acceptable for them to differ if the difference is small enough to be unimportant in comparison to sampling errors, or if the difference can be adjusted for in some way. For example, Wilson (1959) showed that 84% of the eggs of the spruce budworm *Choristoneura fumiferana* in Minnesota (the target population) are laid on the tips of branches. If the sampled population is eggs laid on tips of branches then much effort may be saved at the expense of a small, fairly constant undercount.

The sampled population will consist of a set of individual items, each of which has a numerical value for a variable Y of interest. For example, an item might be a leaf, with Y being the number of insects on the leaf. The **sampling unit** will then either be a single leaf or a collection of leaves. It is important that each leaf in the sampled population is associated with one and only one sampling unit. A **sampling frame**, if one exists, is a list of all the sampling units in the population. If the sample units are plots of ground then a map of the area covered by the plots is a type of sampling frame.

2.3   SIMPLE RANDOM SAMPLING

A **simple random sample** of size n items is one drawn from a population in such a way that every item in this population has the same probability of being included in the sample. What is important here is the process of selection rather than the outcome. Thus, a sample may be random even though all the items in it happen by chance to come from a small part of the population. For example, a random sample of the plots in a field might be such that considerably more than half of the plots come from the left-hand side. This in itself does not mean that the sample is invalid.

Sampling can be **with replacement**, in which case sampled items are replaced in the population and may be included in the sample more than once, or **without replacement**, in which case items are not allowed to be sampled more than once.

If the sampling units in a population can be numbered from 1 to N then one way to draw a random sample involves selecting n of these numbers using a table of random numbers such as the one shown in Table 2.1. This is a table constructed in such a way that each of the digits 0–9 was equally likely to occur in any position, independent of what occurred in any other position.

As an example of the use of Table 2.1, suppose that the trees in an orchard are numbered from 1 to 650 and a random sample of 10 of these trees is required to be taken without replacement. Starting arbitrarily in row five of the table gives the following sequence of four-digit numbers: 8524 5391 9662 7623 0393 1120 1686 4258 5698 0501 . . . . Random sampling can be carried out by taking the first three of each of these sets of four digits and choosing the corresponding tree for the sample, providing that the number is between 1 and 650, and has not been used before. On this basis a random sample of 10 trees consists of those numbered 539, 39, 112, 168, 425, 569, 50, 16, 573 and 167. Some computer programs will also make this type of random selection.

If $\bar{y}$ is the mean of a random sample of n units taken without replacement

Table 2.1　A table of random numbers

6583	1284	0661	6656	8696	9946	1823	1917	8947	0550
7751	0688	8114	0266	4789	8911	2609	0362	7546	6667
7757	5218	8166	0448	3511	9106	7202	5606	9356	6274
6335	2909	7940	3664	3264	4828	8171	2459	8741	4005
8524	5391	9662	7623	0393	1120	1686	4258	5698	0501
8926	0168	5737	1679	1898	9011	7841	9452	8089	4935
6576	3044	5566	8427	7894	3429	6942	4738	3837	9375
3326	3568	6289	0238	1512	2524	7791	9596	5046	7201
6880	0126	3501	0470	5096	9538	7209	4271	1594	4855
8554	0173	9949	6341	6371	8386	7325	6542	3358	1881
2969	7833	9889	3339	3987	2283	6796	0637	8289	4587
4233	6690	3086	2140	8373	2818	6105	0542	1047	7114
9936	4664	1578	1421	8873	9811	7117	7280	8673	8430
3552	7122	7368	4274	2235	5893	5845	5428	3962	5554
0150	4376	1534	0000	4499	1226	2272	4357	2344	3966
6491	6241	8870	8233	0419	5195	1386	9739	7484	4668
1814	1964	7174	4200	7112	2583	5263	2210	5730	2961
1471	8076	1837	3422	5236	3103	2133	1014	6785	5704
7659	4764	3789	7254	5787	1311	1697	4120	2743	6578
7819	4314	4748	5197	9459	6266	2301	0019	9712	7255
4376	9049	6834	7082	3098	1317	2603	3572	0426	2265
9053	7219	8771	0905	6565	0170	8086	2140	8269	8414
6044	8115	8668	6268	7806	0166	5739	5894	0413	7807
3511	5989	6342	8641	2984	3231	7184	7024	0660	5190
4815	9124	5889	3493	5690	6851	3758	7948	3940	1109
5096	9990	1160	5525	8096	9051	1817	0366	6385	7782
2778	0829	5679	2320	1204	2031	0809	0905	0630	6045
8269	7719	5865	7599	8833	3069	6137	4179	7701	6075
8127	6012	1846	9156	0336	5217	4871	2949	7393	8288
8723	2241	0811	1381	5194	6429	3135	9538	2371	1986
9857	9074	6508	4284	7053	3463	4302	9282	6756	8504
9932	2824	2182	8472	6430	6028	0795	9321	0795	0653
6206	3873	9847	1230	0136	6673	1984	9114	1759	6845
6636	4232	0931	4927	5720	1172	2556	3493	5847	3573
6055	7731	1070	8548	7176	3947	2163	1569	0502	0352
4912	0217	1190	7037	9443	4510	9100	6564	1717	6244
7757	7676	6939	7426	6168	6429	1648	7707	5475	6928
5470	1841	8644	0326	6302	3677	7428	1744	9390	6325
9311	2505	2914	6564	5530	3102	1730	2318	4197	3900
7728	4917	1363	9605	7197	0966	4932	9508	6749	6377

then this is an unbiased estimator of μ, the mean of the population of N items, with variance

$$\text{var}(\bar{y}) = (\sigma^2/n)(1 - f). \tag{2.1}$$

Here, $f = n/N$ is called the **finite population correction**. If sampling is with replacement then the finite population correction does not apply and $\text{var}(\bar{y}) = \sigma^2/n$. Hence, sampling without replacement gives a smaller variance than sampling with replacement, although the difference is negligible if N is much larger than n.

The population variance σ^2 can be estimated by the usual formula for the sample variance

$$s^2 = \sum_{i=1}^{n} (y_i - \bar{y})^2/(n-1) = \left\{ \sum_{i=1}^{n} y_i^2 - \left(\sum_{i=1}^{n} y_i \right)^2 / n \right\}/(n-1),$$

where $y_1, y_2, \ldots, y_n$ are the n sample values. The estimated **standard error** of $\bar{y}$ (the square root of the variance) is then

$$\text{SE}(\bar{y}) = (s/\sqrt{n})\sqrt{(1 - f)} \tag{2.2}$$

for sampling without replacement. For sampling with replacement the factor $\sqrt{(1 - f)}$ is omitted.

If the sample size is reasonably large (say 25 or more) and N is 100 or more, $\bar{y}$ will be approximately normally distributed for samples from many distributions met in practice. For this reason, it is common to take

$$\bar{y} - 2\text{SE}(\bar{y}) < \mu < \bar{y} + 2\text{SE}(\bar{y}) \tag{2.3}$$

as an approximate 95% **confidence interval** for the true population mean. Some people use 1.96 instead of 2 for this interval, 1.96 being the correct value for a normal distribution with $\text{SE}(\bar{y})$ known exactly. Use of the factor 2 gives slightly wider limits to compensate for slight non-normality and the fact that the standard error is only estimated.

Many surveys are concerned with population totals rather than with means. For example, if the sampling units are plots of ground, and the variable recorded is the count of the number of insects found on each plot, then the main interest may be T, the total number over the whole area surveyed. This is estimated by $t = N\bar{y}$, N being the total number of plots in the area. The standard error of t is estimated by

$$\text{SE}(t) = N\,\text{SE}(\bar{y}), \tag{2.4}$$

with an approximate 95% confidence interval for T being

$$t - 2\text{SE}(t) < T < t + 2\text{SE}(t). \tag{2.5}$$

Another possibility is that the population proportion π is of interest. An example of this would be where the sampling units are individual animals and π is the proportion of the population with lice. In this case, if a random sample of n animals from the population gives r with lice then the sample proportion $p = r/n$ is an unbiased estimator of π with standard error

$$SE(p) = \sqrt{[\{\pi(1 - \pi)/n\}\{1 - f\}]}, \qquad (2.6)$$

assuming sampling without replacement. For sampling with replacement the factor $(1 - f)$ is omitted. Because π is unknown this standard error must be estimated in practice by

$$SE(p) = \sqrt{[\{p(1 - p)/n\}\{1 - f\}]}. \qquad (2.7)$$

An approximate 95% confidence interval is then

$$p - 2SE(p) < \pi < p + 2SE(p). \qquad (2.8)$$

These equations for estimating a proportion only apply if p is a proportion of sampling units. If each sampling unit has a proportion measured on it (for example, the measurement is the proportion of a sample plot covered with grass) then this variable should just be treated as a measured or counted variable Y and analysed using equations (2.2) and (2.3).

Example 2.1   Counting Fir Trees in a Forest

Suppose that the number of fir trees in a forest is to be estimated as part of a study of insect damage. The forest contains 6000 plots of size 0.1 ha, from which a simple random sample of 25 plots is chosen. The number of fir trees is recorded for each of the sampled plots, as well as an indication of whether or not any trees on the plot show severe damage by the insect. Results are by follows:

Plot:	1	2	3	4	5	6	7	8	9	10	11	12	13	14
Number of trees:	3	3	2	1	1	5	3	4	2	2	6	1	2	1
Damage (yes or no):	y	y	n	n	n	y	n	n	y	n	y	n	n	n

Plot:	15	16	17	18	19	20	21	22	23	24	25
Number of trees:	5	3	3	2	1	1	1	3	2	4	2
Damage (yes or no):	n	y	y	n	n	n	y	n	n	y	n

Turning first to the estimation of the total number of trees in the forest, the sample counts have a mean of $\bar{y} = \Sigma y_i/n = 63/25 = 2.52$, and a variance of $s^2 = \{\Sigma y_i^2 - (\Sigma y_i)^2/n\}/(n - 1) = \{207 - 63^2/25\}/(25 - 1) = 2.01$. The standard error of $\bar{y}$ is therefore estimated from equation (2.2) as $SE(\bar{y}) = \sqrt{\{(2.01/25)(1 - 25/6000)\}} = 0.283$. The estimated total number of trees in the forest is $N\bar{y} = 6000 \times 2.52 = 15\,120$, with standard error $N.SE(\bar{y}) = 6000 \times 0.283 = 1698$. Approximate 95% confidence limits for the true total are given by equation (2.5) to be $15\,120 - 2(1698) < T < 15\,120 + 2(1698)$ or $11\,724 < T < 18\,516$.

Nine out of the 25 sampled plots showed at least one severely damaged tree. The estimate of the proportion of all plots with this characteristic is therefore $p = 9/25 = 0.36$. From equations (2.7) and (2.8) the estimated standard error associated with this estimate is $SE(p) = \sqrt{[\{0.36(1 - 0.36)/25\}(1 - 25/6000)]} = 0.096$, and the approximate 95% confidence limits for the population proportion are $0.36 - 2 \times 0.096 < \pi < 0.36 + 2 \times 0.096$ or $0.17 < \pi < 0.55$. The large range reflects the very small sample size for estimating a proportion.

2.4 DETERMINING SAMPLE SIZES WITH SIMPLE RANDOM SAMPLING

An important question for the design of a sampling programme is the sample size required to achieve a particular level of precision. This can be determined by setting the confidence bounds to reflect this level. Thus, to get a 95% confidence interval for a population mean of $\bar{y} \pm \delta$, with δ specified, requires $\delta = 2(\sigma/\sqrt{n})\sqrt{(1 - n/N)}$. Solving for n gives

$$n = N\sigma^2/(N\delta^2/4 + \sigma^2). \tag{2.9}$$

For large population sizes N this becomes

$$n = 4\sigma^2/\delta^2, \tag{2.10}$$

this being the equation to use if N is large and unknown.

These equations for sample size require σ to be known. Sometimes an approximate value is available from a previous study. Alternatively, if the range of possible values of Y is guessed to be R then $R/4$ will give an approximate σ, the idea being that for many distributions the effective range is the mean plus and minus about two standard deviations. This type of approximation is often sufficient in practice because it is only necessary to get a rough idea of the appropriate sample size.

To obtain specified bounds on a population total, these bounds can be expressed as bounds on the mean and equation (2.9) can be used. Thus if the required 95% confidence interval for T is $t \pm d$ then this can be written as $N\bar{y} \pm N\delta$. The required interval for the population mean is then $\bar{y} \pm \delta$, where $\delta = d/N$.

To obtain specified bounds on a population proportion, say an approximate 95% confidence interval of $p \pm \delta$, requires that $\delta = 2\sqrt{[\{\pi(1 - \pi)/n\}\{1 - n/N\}]}$. Solving for n then gives

$$n = N\pi(1 - \pi)/\{N\delta^2/4 + \pi(1 - \pi)\}. \tag{2.11}$$

To use this equation a value for π is needed. This value for π can be guessed, or a conservative value can be used. The conservative approach involves noting that the largest possible sample size is needed when $\pi = \frac{1}{2}$, in which case equation (2.11) becomes

$$n = N/(N\delta^2 + 1). \tag{2.12}$$

If the population size N is large this gives

$$n = 1/\delta^2. \tag{2.13}$$

Equation (2.13) is useful as a general guide for sample sizes for surveys. Whenever samples are to be used to estimate proportions, it can be seen that to obtain an accuracy of about ±0.1 needs a sample size of $n = 1/0.1^2 = 100$, an accuracy of ±0.05 needs $n = 1/0.05^2 = 400$, an accuracy of ±0.01 needs a sample size of $n = 1/0.01^2 = 10\,000$, etc.

2.5   STRATIFIED RANDOM SAMPLING

A drawback with simple random sampling that was mentioned earlier is that there is no control over how sampled items are spread through a population. Thus, in sampling plots of land in a field the sampled plots may be concentrated mostly in a small area rather than being well spread out. This does not matter if the variable being studied has a fairly uniform distribution over the field. It is important if plots wth similar values tend to be clustered together.

One way of ensuring that a population is well represented by sampling involves using stratification. This means dividing the sampling units into non-overlapping strata, and selecting a simple random sample from each of these. In general, there is nothing to lose by using this more complicated type of sampling but there are some potential gains. First, if the individuals within strata are rather more similar than individuals in general then the estimate of the overall population mean will have a smaller standard error than can be obtained with a simple random sample size of the same size. Second, there may be value in having separate estimates of population parameters for the different strata. Third, stratification makes it possible to sample different parts of a population in different ways, which may reduce some sampling cost.

In sampling animals and plants the types of stratification that should be considered are those based on spatial location, areas within which the habitat is rather uniform, and the size of sampling units. For example, in sampling an animal population over a large area it is natural to take a map and partition the area into a few relatively homogeneous subregions based on factors such as altitude and vegetation type. These are then the strata. On the other hand, in sampling insects on trees it may make sense to stratify on the basis of the tree diameter. The strata might then be small, medium and large trees. Usually, the choice of what to stratify on is a question of common sense, being based on factors that the biologist feels are likely to affect the magnitude of the variable being studied.

Assume that K strata have been chosen, with the ith of these having size N_i and the total population size being $\Sigma N_i = N$. Then if a random sample with size n_i is taken from the ith stratum the sample mean $\bar{y}_i$ will be an unbiased estimate of the true stratum mean μ_i with estimated variance given from equation (2.2) to be

$$\text{var}(\bar{y}_i) = (s_i^2/n_i)(1 - n_i/N_i), \tag{2.14}$$

where s_i is the standard deviation of Y within the stratum. In terms of the true strata means, the overall population mean is the weighted average

$$\mu = \sum_{i=1}^{K} N_i \mu_i/N. \tag{2.15}$$

The corresponding sample estimate is

$$\bar{y} = \sum_{i=1}^{K} N_i \bar{y}_i/N, \tag{2.16}$$

with estimated variance

$$\text{var}(\bar{y}) = \sum_{i=1}^{K} N_i^2 \, \text{var}(\bar{y}_i)/N^2. \tag{2.17}$$

The standard error of $\bar{y}$ is the square root of this variance and an approximate 95% confidence interval is given by equation (2.3).

If the population total is of interest then this can be estimated by $t = N\bar{y}$. Equations (2.4) and (2.5) can then be used to determine the standard error and confidence limits.

Stratified sampling can also be used with the estimation of proportions. If p_i is the sample proportion in stratum i then this is an unbiased estimator of the stratum proportion π_i with estimated variance

$$\text{var}(p_i) = \{p_i(1 - p_i)/n_i\}\{1 - n_i/N_i\} \tag{2.18}$$

An unbiased estimator of the overall population proportion π is then

$$p = \sum_{i=1}^{K} N_i p_i/N \tag{2.19}$$

with estimated variance

$$\text{var}(\bar{p}) = \sum_{i=1}^{K} N_i^2 \, \text{var}(p_i)/N^2. \tag{2.20}$$

A 95% confidence interval for π is given by equation (2.8), SE(p) being taken as the square root of the estimated variance.

If the sample sizes from the different strata are taken in proportion to the strata sizes then this is called **stratification with proportional allocation**. The samples are self-weighting in the sense that the estimates of the overall mean and the overall proportion given by equations (2.16) and (2.19) are the same as what is obtained by lumping the results from all the strata together as a single sample. However, the stratified sampling variance is not the same as the variance of the simple random sample.

Although proportional allocation is often used because it is most convenient, it is not necessarily the most efficient use of resources. Equations are available for determining sample sizes to achieve a given level of precision for overall population estimates, or to gain the maximum precision at least cost (Cochran, 1977, p. 96). The calculations require approximate values for strata variances and a knowledge of sampling costs. If these variances and costs are about the same in all strata then proportional allocation is optimal.

Example 2.2   Stratified Sampling of an Insect Population

Table 2.2 shows counts of the number of insects in a population covering 1600 small plots, in four strata. This is an artificial example but it will serve to illustrate the potential value of stratified sampling. A comparison will be made between taking a simple random sample of 40 plots, and taking a stratified sample of the same total size comprising 10 from each of the strata.

A simple random sample of 40 was taken by choosing a random row between 1 and 40, and a random column between 1 and 40, this being repeated 40 times. The results were as follows:

Row	Col	Y	Row	Col	Y	Row	Col	Y	Row	Col	Y
29	22	7	15	3	7	32	2	3	13	38	5
11	29	4	4	14	9	8	34	2	9	31	1
8	18	8	35	35	4	32	22	4	38	36	6
3	15	6	6	25	5	20	40	3	3	32	5
35	40	4	36	28	3	13	9	8	14	32	4
22	1	7	22	26	6	2	5	8	7	14	7
27	19	3	21	4	9	13	5	8	11	5	8
1	4	5	39	22	5	18	19	7	4	14	9
10	16	8	18	33	4	34	33	5	40	26	5
30	12	6	17	16	7	33	32	4	32	28	6

Here the sample mean is $\bar{y} = 5.625$, with standard deviation $s = 2.047$. The standard error of the mean is estimated from equation (2.2) to be 0.324. It

Table 2.2 Counts for an artificial population covering the area of 1600 small plots. The total population size is 8530, with a mean of 5.33 individuals per plot

Stratum 1	Stratum 2
77859878886777886742	42155655646446836564
88898784878886867825	27452342465664465454
78587676684966667766	56257436346533564546
77868977958869769834	54542455554573254734
86876878767967865855	63758655742543633576
78798679977678769656	54625445243544415546
87788977887687896745	53227554467465756665
77786689577978768843	77532855554442355734
75967687796777768635	77343426321654386543
88777798886989886832	63454437455742555242
89978776689897788945	48678582433453273254
68776767888886888534	55561322543552765434
65898779854896679854	24666426542454543557
78767887967677657735	74646563557435543435
86767996987978787654	36464635575235535634
56766866898966798936	67644241457555467436
99898778777677776664	43544434355655423253
88597887598876884871	54653435444547537655
97877677755798696976	34524354661522544572
88767879886968896746	53435455563442854343

Stratum 3	Stratum 4
78898568778887877865	24445345343363475464
76758669976795888741	53443663834624524552
57758848977898975964	55326254563436435553
97769788876976679944	46526345425453657436
66786779886687787773	48225342543376553515
78663887477688776645	48567755334535726446
98784676767678775633	54711424334556365634
66787888877685965854	46436334466854553247
86856698877759756836	47515458824747236535
78868866677698788654	26465556645424334492
41322463336638247464	45425365644235556325
03324345245346453255	54454366324646144273
42333223324324652563	45523562544461448476
42222324145633463524	57455436524553486564
76645666535526555545	27664754764451444764
54427556446246332343	46736533525443256634
57736425233657232392	42464665452537464654
74656554775535665625	45345545550337161506
35466345663362366534	85343254635456344675
65245742243525336361	63375534363432455723

follows that the estimated population total is $5.625 \times 1600 = 9000.0$, with standard error $0.324 \times 1600 = 517.8$. An approximate 95% confidence interval for the true total is therefore $9000 \pm 2 \times 517.8$, or 7964 to 10 036.

For stratified sampling, a random row was chosen between 1 and 20, and also a random column in the same range. This was repeated 10 times for each of the four strata with the following results:

Stratum 1			Stratum 2			Stratum 3			Stratum 4		
Row	Col	Y	Row	Col	Y	Row	Col	Y	Row	Col	Y
5	1	8	13	17	3	6	6	8	18	19	0
8	6	6	1	4	5	20	11	3	2	11	4
1	8	8	2	4	5	17	1	5	3	3	3
2	16	6	14	7	6	6	6	8	19	8	4
8	1	7	13	7	2	11	4	2	14	18	5
10	3	7	1	13	4	16	13	4	11	3	4
19	9	7	9	8	6	9	10	7	17	7	6
3	15	6	18	11	4	8	16	6	8	17	3
20	9	8	15	17	5	14	18	5	14	7	3
10	16	8	12	19	3	15	1	7	6	4	4

The following strata means, standard deviations and standard errors are obtained from these samples: $\bar{y}_1 = 7.1$, $\bar{y}_2 = 4.3$, $\bar{y}_3 = 5.5$, $\bar{y}_4 = 3.4$, $s_1 = 0.88$, $s_2 = 1.34$, $s_3 = 2.07$, $s_4 = 1.65$, $SE(\bar{y}_1) = 0.27$, $SE(\bar{y}_2) = 0.41$, $SE(\bar{y}_3) = 0.64$ and $SE(\bar{y}_4) = 0.51$. From equation (2.16) the estimate of the overall population mean is $\bar{y} = (400 \times 7.1 + 400 \times 4.3 + 400 \times 5.5 + 400 \times 3.4)/1600 = 5.075$, which is just a simple average of the strata means. The variance of $\bar{y}$ is found from equation (2.17) to be $\text{var}(\bar{y}) = (400^2 0.27^2 + 400^2 0.41^2 + 400^2 0.64^2 + 400^2 0.51^2)/1600^2 = 0.057$. The standard error of the estimator is therefore $\sqrt{0.057} = 0.238$, which compares well with the value of 0.324 for simple random sampling.

The estimated population total is $5.075 \times 1600 = 8120$, with estimated standard error $0.238 \times 1600 = 380.8$. The approximate 95% confidence interval for the true value is $8120 \pm 2 \times 380.8$, or 7359 to 8882. The estimate is closer to the true value than the estimate from simple random sampling and the confidence interval is narrower with stratified sampling. This is quite understandable because the population density clearly varies from stratum to stratum in the population, with relatively high counts in stratum 1 and part of stratum 3.

2.6   RATIO ESTIMATION

Occasions arise where the estimation of the population mean or total for a variable Y is assisted by information on a subsidiary variable U. For example, suppose that an estimate is required of the total number of insects of a certain species in an area divided into N plots. Suppose also that two types of insect count can then be made on each of these plots: (1) a quick sweep can be made with a net to get a count U that will be too low, and (2) a time-consuming search can be made to get an accurate count Y. One possibility is then to do the quick sweep on all plots and hence determine the population total T_u of U, and make the slow search on a random sample of n plots to determine accurate counts $y_1, y_2, \ldots, y_n$, with mean $\bar{y}$. On the assumption that the ratio of Y to U is fairly constant from plot to plot, this ratio can then be estimated by

$$r = \bar{y}/\bar{u}, \tag{2.21}$$

where $\bar{u}$ is the mean of the U values $u_1, u_2, \ldots, u_n$ on the n randomly sampled plots. The ratio estimator of the Y total T_y for the whole field is then

$$t_y = rT_u. \tag{2.22}$$

This can be expected to have a lower standard error than the estimator $N\bar{y}$ that uses only the Y values as T_y adjusts for the fact that the random sample may, by chance alone, consist of plots with rather low or high counts. Thus, even if the random sample does not reflect the population very well the estimate of r, and hence T_y, may still be reasonable.

The variance of t_y can be estimated from the equation

$$\text{var}(t_y) = \{N^2/n\}\left\{ \sum_{i=1}^{n} (y_i - ru_i)^2/(n-1)\right\}(1 - n/N). \tag{2.23}$$

As before, an approximate 95% confidence interval for the population Y total is given by $t_y \pm 2\text{SE}(t_y)$, the standard error being the square root of the variance. These results require the sample size n to be 'large'. Cochran (1977, p. 153) suggests that this requires $n > 30$ and the coefficients of variation of $\bar{y}$ and $\bar{u}$ (their standard deviations over their means) to be less than 0.1.

If it is the population mean of Y, μ_y, that is of interest then this is obtained by dividing equation (2.22) by N to get the estimator

$$\bar{y}_R = r\mu_u. \tag{2.24}$$

This has a standard error that can be estimated by

$$\text{SE}(\bar{y}_R) = \text{SE}(t_y)/N. \tag{2.25}$$

An approximate 95% confidence interval for μ_y is $\bar{y}_R \pm 2\text{SE}(\bar{y}_R)$. Again, the variance and confidence intervals are large sample results that should work well under the same conditions as the results for the mean.

Ratio estimators are not unbiased, although the bias will be unimportant for large samples. Cochran (1977, p. 162) notes that for estimation of means or totals the bias can be bounded by the equation

$$|\text{bias}| \leq \text{SE}(\bar{u})\,\text{SE}(\text{estimate})/\mu_u. \qquad (2.26)$$

This equation can be used to check that the bias is not potentially serious.

It is possible to design a sample using ratio estimation in order to achieve a predetermined level of accuracy at the least cost, or the best possible accuracy for a fixed cost. Formulae are provided by Cochran (1977, p. 172).

Ratio estimation can be combined with stratified sampling in two ways. It can be used separately in each of the strata using the formulae given in this section. Equations (2.16) and (2.17) can then be used to determine an overall population mean with a standard error. This **separate ratio estimation** is appropriate, providing that the sample sizes in all strata are large enough (say 20 or more) to estimate ratios well. Alternatively, the means of U and Y can be determined using the usual stratified sampling formulae and the ratio of these, r, used as the estimator of μ_y/μ_u for equations (2.22) and (2.24). This **combined ratio estimation** is appropriate with small sample sizes in strata, or if there is reason to believe that the ratio of U to Y is nearly the same in all strata.

Example 2.3   Sampling Trees in a Forest

Suppose that the total number of insects in a forest is to be estimated. These insects occur only in trees, so that a tree is the natural sampling unit. However, it is easier to divide the forest area into 240 plots, containing variable numbers of trees, and use the plots as the sampling unit. The insect count on a plot should then be approximately proportional to the number of trees. From an aerial photograph the total number of trees in the forest is known to be 1742. Ratio estimation is therefore appropriate with the number of trees per plot as the subsidiary variable U.

Table 2.3 shows a model population for this situation. A simple random sample of 20 was taken from this and the U and Y values shown in Table 2.4 were obtained. A plot of Y against U confirms that these are approximately proportional (Figure 2.1), with the plotted points being distributed about a line through the origin.

The mean and standard deviation of the U values are $\bar{u} = 7.25$ and $s_u = 2.53$. For the Y values these statistics are $\bar{y} = 101.7$ and $s_y = 48.50$. The ratio of Y to U is estimated as $r = 101.7/7.25 = 14.03$. The ratio estimator of the Y total is therefore $t_y = 14.03(1742) = 24\,440$ from equation (2.22). From equation (2.23) the standard error of t_y is estimated to be 2062.3, so the approximate 95% confidence interval for the total insect

Table 2.3 A model population for ratio sampling with U = number of trees and Y = the number of insects on the trees for a forest consisting of 240 plots in six rows of 40 plots. The population totals for U and Y are T_u = 1742 and T_y = 23 902 and the means are μ_u = 7.26 and μ_y = 99.6

	1		2		3		4		5		6	
	U	Y	U	Y	U	Y	U	Y	U	Y	U	Y
1	10	46	8	138	8	47	11	167	10	106	9	121
2	10	161	10	79	11	108	6	111	2	45	5	150
3	12	54	6	83	7	121	2	41	3	54	3	60
4	5	76	9	173	7	166	9	35	8	87	11	185
5	5	70	8	11	12	93	7	97	8	96	7	6
6	10	89	9	163	6	83	10	52	6	78	5	114
7	10	123	3	56	9	104	8	36	7	76	2	56
8	5	74	8	76	10	147	8	37	8	90	10	124
9	7	44	7	57	11	29	6	114	10	65	9	83
10	9	67	5	59	5	22	9	106	9	175	8	76
11	5	65	6	71	10	134	3	31	9	76	10	146
12	8	129	6	135	11	115	16	138	6	47	5	49
13	5	5	8	213	15	228	8	125	4	0	10	147
14	6	53	10	107	11	181	9	148	9	115	5	52
15	7	2	9	132	9	154	6	63	10	63	11	183
16	8	86	4	52	12	123	11	68	5	50	6	48
17	6	150	7	131	12	84	9	87	6	137	9	77
18	4	42	5	51	7	120	7	100	7	44	9	67
19	6	64	3	75	7	53	7	89	8	163	8	125
20	7	90	12	306	9	135	11	120	5	96	8	128
21	4	90	8	123	8	133	6	151	9	82	6	8
22	7	140	6	80	5	93	4	79	7	131	7	73
23	7	25	4	63	10	66	6	100	5	88	3	66
24	5	14	7	144	5	18	5	84	11	198	9	105
25	8	43	8	56	7	22	6	172	6	64	8	182
26	6	59	3	59	6	157	4	90	7	65	7	44
27	9	105	5	114	4	89	7	108	6	125	5	73
28	11	186	9	155	5	73	4	70	7	55	6	149
29	2	68	9	162	6	63	10	123	7	90	3	80
30	8	99	6	80	9	16	13	270	8	88	4	60
31	8	120	7	131	5	139	10	106	7	157	5	110
32	6	122	12	131	11	204	7	167	9	166	3	50
33	9	181	5	68	10	98	10	198	6	80	8	91
34	9	173	11	181	7	158	4	51	9	131	9	100
35	8	108	7	142	12	116	5	95	12	228	4	129
36	6	63	8	138	7	45	8	111	8	89	9	145
37	8	167	4	80	7	72	5	113	4	101	1	71
38	6	127	6	95	3	60	3	103	4	80	7	119
39	11	74	3	40	7	110	3	71	7	71	7	100
40	8	126	1	55	10	170	6	127	8	96	9	154

Table 2.4 A random sample of 20 plots from the population in Table 2.3. The values of rows and columns indicate where the sample points are in the population

Row	Column	U	Y	Row	Column	U	Y
1	2	8	138	3	4	2	41
3	6	3	60	4	2	9	173
10	4	9	106	10	6	8	76
11	2	6	71	12	2	6	135
12	5	6	47	16	3	12	123
18	5	7	44	19	5	8	163
20	3	9	135	22	6	7	73
24	3	5	18	25	5	6	64
25	6	8	182	31	6	5	110
32	2	12	131	36	6	9	145

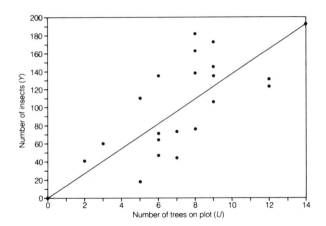

Figure 2.1 A plot of sample *Y* values against *U* values that confirms that the number of insects on a plot is approximately proportional to the number of trees on the plot.

population is $24\,440 \pm 2(2062.3)$, or 20 315 to 28 565. The actual population total is 23 902.

   If the *U* values were not known then the sample of *Y* values would have to be analysed using the equations (2.2) to (2.5) for simple random sampling. The estimated population total is then $t = N\bar{y} = 240 \times 101.7 = 24\,408$, with estimated standard error $SE(t) = N.SE(\bar{y}) = 240 \times 48.50/\sqrt{20} = 2602.8$. This is 26% higher than the estimated standard error of the ratio estimator.

2.7   REGRESSION ESTIMATION

Ratio estimation assumes that the ratio of the variable of interest Y to the subsidiary variable U is approximately constant from plot to plot. A less restrictive assumption is that Y and U are approximately linearly related by a regression equation $Y = a + bU$. Regression estimators of the mean and total of Y are then available.

Suppose that a simple random sample of n is taken from a population and yields Y values $y_1, y_2, \ldots, y_n$, and corresponding U values $u_1, u_2, \ldots, u_n$. Then the standard equations for the regression coefficients a and b are

$$b = \sum_{i=1}^{n} (y_i - \bar{y})(u_i - \bar{u}) / \sum_{i=1}^{n} (u_i - \bar{u})^2 \tag{2.27}$$

and

$$a = \bar{y} - b\bar{u}, \tag{2.28}$$

where $\bar{y}$ is the sample mean of Y and $\bar{u}$ is the sample mean of U. The linear regression estimator of the mean of Y, μ_y, is then

$$\bar{y}_L = a + b\mu_u = \bar{y} + b(\mu_u - \bar{u}), \tag{2.29}$$

which can be interpreted as $\bar{y}$ corrected by $b(\mu_u - \bar{u})$ for a low or high mean of the U values for the sampled items.

The variance of $\bar{y}_L$ can be estimated by

$$\mathrm{var}(\bar{y}_L) = \left\{ \sum_{i=1}^{n} (y_i - \bar{y})^2 - b^2 \sum_{i=1}^{n} (u_i - \bar{u})^2 \right\} / \{(n-2)n\}(1 - n/N), \tag{2.30}$$

and an approximate 95% confidence interval for μ_y is $\bar{y}_L \pm 2\mathrm{SE}(\bar{y}_L)$.

Multiplying equation (2.29) through by the population size N gives the regression estimator of the population total of Y, T_y, to be

$$t_y = N\bar{y} + b(T_u - N\bar{u}). \tag{2.31}$$

The standard error is $\mathrm{SE}(t_y) = N.\mathrm{SE}(\bar{y}_L)$ and approximate confidence limits for T_y are $t_y \pm 2\mathrm{SE}(t_y)$.

Example 2.4   Sampling Trees in a Forest, Revisited

The data given in Table 2.4 for a random sample of 20 plots from the 240 plots of trees in a forest can be used to determine a regression estimate of the total number of insects in the forest. Actually, in this example it is clear that the number of insects on a plot can be expected to be proportional to the number of trees on the plot (U), so that ratio estimation is definitely appropriate. Nevertheless, this can be used as an example of regression estimation since a ratio relationship between U and Y is a special case of a regression relationship. Using regression estimation when ratio estimation is appropriate will generally just result in slightly larger standard errors than would be obtained with ratio estimation.

The estimate of b from equation (2.27) is 11.08. The regression estimate of the population total of the Y values from equation (2.31) is $t_y = 24\ 430$, with standard error = 2142. Approximate 95% confidence limits for the true total are 20 146 to 28 714. These results are very similar to what was obtained for ratio estimation in Example 2.3.

2.8   CLUSTER SAMPLING

Cluster sampling involves each sampling unit being a group or cluster of items rather than a single item. Cost reduction is the usual reason for using this type of sampling since it often costs much less to sample several close items than the same number of items some distance apart. Examples 2.3 and 2.4, concerning sampling trees in a forest, can be thought of as cluster sampling of trees rather than sampling of plots.

Generally, cluster sampling cannot be expected to give the same precision as a simple random sample with the same total number of items. This is because it is usually the case that close items tend to be more similar than items in general. Therefore, a cluster sample is equivalent in effect to a random sample of independent units with a somewhat smaller size. However, cost savings may allow a cluster sample to be considerably larger than a simple random sample could be. Hence, the cluster sample may give better precision for the same sampling cost.

If cluster sizes are constant at m then the simplest way to estimate the mean of the variable of interest from a random sample of n clusters involves defining y_i as the total of the observations in the ith sampled cluster so that $\bar{y}$ is an unbiased estimator of the population mean for cluster totals. Equations (2.2) and (2.3) can then be used to estimate the standard error of $\bar{y}$ and put 95% confidence limits on the population mean. The mean for individual items is estimated by $\bar{y}/m$, with 95% confidence limits being obtained by dividing the limits $\bar{y} \pm 2\mathrm{SE}(\bar{y})$ by m.

If cluster sizes vary then a cluster total Y can be expected to be approximately proportional to the cluster size. This size can therefore be used as the value for the subsidiary variable U for ratio estimation and the formulae given in Section 2.6 can be used, taking N equal to the number of clusters in the population. Essentially this was the procedure used in Example 2.3.

2.9   SYSTEMATIC SAMPLING

Systematic sampling can be carried out whenever a population can be listed in order or it covers a well-defined spatial area. In the former case, every kth item in the list can be sampled, starting at an item chosen at random from the first k. In the second case, sampling points can be set out on a grid at equally spaced intervals.

Systematic sampling may give a better coverage of a population than a simple random sample. However, it suffers from the disadvantage of not allowing any determination of the level of sampling errors unless it is assumed that the items in the population are in a more or less random order. If that is the case then a systematic sample can be treated as being effectively a simple random sample and the various results given earlier for this type of sampling can all be used.

Another possibility involves the estimation of a sampling variance by replicating a systematic sample. For example, suppose that a 10% sample of a population is required. Rather than taking every 10th item, starting with one of the items 1–10, randomly chosen, it might be possible to take 20 systematic samples, each starting at a different randomly chosen item in the first 200, and sampling every 200th item from then on. The population can then be thought of as consisting of $N = 200$ 'clusters' from which 20 are randomly sampled. Inferences concerning the population mean and total can be made in the manner indicated in the previous section.

2.10   MULTI-STAGE SAMPLING

Sometimes sampling has to be conducted at several levels of the classification of a population. This is called **multi-stage sampling**. For instance, to estimate the number of insects in a forest, the area covered might be stratified into geographical areas, and random samples taken of the trees in each of these areas. It may then not be practical to count all the insects on these trees so one or two branches may be chosen at random from each tree for this purpose. It may then still not be possible to count the insects on all the leaves of a branch so a small number of leaves may be chosen at random for counting. This gives a three-stage sampling design, with random samples taken at the tree level, branch level and leaf level. A specialized text such as Cochran (1977) should be consulted for appropriate estimation procedures for these types of samples.

2.11   SAMPLING FOR STAGE-FREQUENCY DATA

Since much of this book is concerned with the analysis of stage-frequency data, it is appropriate to give some special consideration to the sampling problems involved with collecting data of this type for a wild population. Recall that stage-frequency data consists of counts of the numbers of individuals in different life stages in a population (or a fraction of the population) at a series of points in time. The general form is shown in Table 1.1 and a particular example in Table 1.2.

The simplest way to obtain stage-frequency data involves dividing the area covered by a population into N equal-sized plots, taking a simple random sample of n of these (without replacement), and counting the number in each life stage on each sample plot. For many purposes the plot totals can then be used without any adjustment for the sampling fraction.

If the sampling area is not uniform there may be advantages in using stratification. This can be done on the basis of vegetation and soil type, etc., if information on these is available. Alternatively, an arbitrary division of the area into a few approximately equal-sized subareas may capture most of the important variation simply because any subarea is likely to be more uniform than the total area. The various formulae for stratification given above should then be applied separately to the counts for different life stages.

When the number of plots in the population (N) is much larger than the number of sampled plots (n) it may be reasonable to assume that the number counted on all plots in a particular life stage at one sample time is a random value from a Poisson distribution, with a mean equal to the total number in the population in that life stage times the sampling fraction n/N. If this is true then the sampling variance of the count should equal the expected count. A Poisson distribution for the total sampled in a stage will arise if the individuals in the population are distributed independently, at random over the sampled region. Often, however, individuals will tend to occur in clusters and the obtained sampling variance will be larger than the expected count. For this reason, the sampling variance is best estimated from the variation in the n sampled plots using equation (2.1) whenever this is possible.

In point of fact, many methods for analysing stage-frequency data do not make use of sampling variances for the counts of the numbers in stages. This is something of an advantage to the extent that the methods can still be applied when these variances are unknown. However, in the case of models fitted by maximum likelihood a knowledge of these variances may be useful for assessing the fit of the models. This is because with models fitted by maximum likelihood it is necessary to make assumptions about the distributions of sampling errors. Usually, it is assumed that these follow Poisson distributions. However, as has just been noted, it is common to find that the actual sampling errors are more variable than the Poisson distribution suggests. When it occurs, extraneous sampling variance will tend to show up in large goodness-of-fit statistics and, as will be explained in Section 3.6, these statistics can be used to determine the heterogeneity factor H, which is an estimate of the ratio of the obtained sampling variance to the Poisson sampling variance. On the assumption that this ratio is constant for all the stage-frequencies, the heterogeneity factor can then be used for adjusting the variances of population parameters for the extraneous variance.

The value of having independent estimates of sampling variances from sampling n random plots comes from the fact that these allow an alternative estimation of the heterogeneity factor, and make it possible to check whether this seems to be constant for all samples at all times. Thus, let M and V be the mean and variance, respectively, obtained from n plot counts for one life stage taken at one time. On the Poisson assumption, the ratio V/M should be approximately one. Values greater than one indicate extraneous variance, and should approximately equal the heterogeneity factor obtained by fitting a model by maximum likelihood. Plotting V/M values against the sample time and the life stage will give an indication of whether the extraneous variance is constant.

One way to interpret the use of a heterogeneity factor is in terms of effective sample sizes. For example, if sampling variances are twice what is expected from Poisson errors then it is as if the sampling fraction was only half as much as the nominal level, and all the observed frequencies were then doubled. More generally, a heterogeneity factor of H indicates that effective sample sizes are the observed sample sizes divided by H.

Braner and Hairston (1989) suggest another approach to handling extraneous sampling variance. They assume that sampling success varies from sample to sample according to random values from a gamma distribution with a mean of unity and a shape parameter to be estimated from the data. This has the merit of taking into account the large variation in the effective sampling intensity from day to day that is observed with some data. The data in the example that follows this section is a good illustration of this type of effect.

It is a different matter when the sampling intensity is not the same at all sample times because of the sampling design. For example, it might be found that the sampling fraction in the first few samples taken from a population is insufficient to get reasonably high stage-frequency counts. More effort might then be put into the sampling to correct for this. Controlled variation of this type can be handled either approximately, by adjusting counts to a common sampling fraction or, more accurately, by making an allowance in the estimation process.

A major difficulty with stage-frequency sampling is that there is a tendency for some life stages to be under-counted because of their small size or the behaviour of the individuals. It may be necessary to adjust counts for the biased sampling, doing experiments to estimate the biases involved (McDonald and Manly, 1989; Schneider, 1989). For example, if experiments indicate that about 90% of adults but only about 60% of first instar larvae are counted in samples then the estimated total adult count should be adjusted by dividing by 0.9 and the estimated total first-instar count should be divided by 0.6. Providing that the sampling efficiencies are known accurately for different stages, the standard errors of the adjusted totals are just the standard errors of the unadjusted totals divided by the same correction factors.

Although random sampling of plots to determine stage-frequencies has the advantage of allowing models to be fitted to data assuming a Poisson distribution of sampling errors, it is not necessarily the best scheme in terms of getting good estimates of population parameters. Munholland (1988) carried out a study of a population of the yellow birch lace bug *Corythucha pallipes* that involved both random sampling and the repeated counting of stages in a part of the population of interest. As might be expected, she found the latter data to display far less variation. Indeed, the data from random sampling was considered too erratic to be analysed alone.

It seems, therefore, that a reasonable approach to obtaining data might involve the repeated sampling of a fixed fraction of a population. Estimates can then still be obtained as if there were Poisson sampling errors and the estimates of population parameters should still be realistic. However, simulation would probably be the only reasonable way to determine the accuracy of these estimates.

Example 2.5   Sampling Zooplankton in Teapot Lake

A study described by Rigler and Cooley (1974) gives some idea of the difficulties involved in collecting stage-frequency data for a natural population. This took place in 1966 at Teapot Lake, a 0.5-ha circular lake in southern Ontario. The organism sampled was the planktonic copepod *Skistodiaptomus oregonensis*, which is normally the dominant member of the zooplankton when the lake is not covered by ice in winter.

Plankton samples were collected at depth intervals of 0.5 m using a 10-l Juday trap fitted with a 67-μ Nitex net. Five sampling points were used, evenly spaced across the lake, roughly in the direction of the wind. The trap was closed for each depth as soon as it arrived. After washing and removal of some unwanted organisms, the numbers of *S. oregonensis* were counted for each depth. Either full counts or a sample count were made, depending on the abundance. The average of all samples from one depth was then taken to represent a stratum of water covering a depth of 0.5 m centred on the sampling depth. The mean number of animals in the 10-l samples from a given depth, multiplied by 100, multiplied by the volume of the stratum in the lake in cubic metres, was taken as an estimate of the number in that stratum. The total from all the strata then gave an estimate of the total lake population. Separate counts were made for each of six naupliar stages (N1–N6), five copepodite stages (CI–CV) and the adult stage (see Table 2.5).

The sampling variation seems to have been rather large for this study. For example, the population size is estimated to have increased by 72% between days 40 and 43. However, this does not seem to have been a genuine change since the increase is primarily in the more advanced stages. Obviously, any new entries to the population should show up first in the early stages. It seems that although Rigler and Cooley were careful to maintain a consistent sampling plan, there was variation in the sampling intensity from day to day,

Table 2.5 Estimated population frequencies (millions in the entire lake) for the copepod *Skistodiaptomus oregonensis* in Teapot Lake, Ontario, in 1966 (the population was killed by anaerobiosis shortly after day 228. N1–N6 are naupliar stages, CI to CV are copepodite stages, A indicates adults)

Day	N1	N2	N3	N4	N5	N6	CI	CII	CIII	CIV	CV	A	Total
6	26	4											30
11	81	38											119
16	161	237	45	3									446
19	27	176	182	58	15	1							459
22	33	68	192	330	258	46	2						929
25	25	55	68	119	243	311	94	14					929
28	37	51	53	68	112	205	203	30	2				761
32	16	54	54	51	84	165	177	102	47	17	3		770
36	18	49	41	51	45	94	97	90	105	104	48	1	743
40	9	16	37	23	49	63	98	64	96	113	122	9	699
43	17	23	49	41	40	82	110	121	160	173	248	138	1202
47	37	16	13	25	44	54	62	57	96	115	101	301	921
50	78	145	41	12	16	19	33	44	67	78	53	172	758
57	181	437	583	168	69	30	15	11	52	101	54	196	1897
64	111	186	294	238	309	455	289	56	42	83	50	201	2314
71	14	15	84	144	125	142	222	322	126	157	88	255	1694
77	17	46	23	39	29	64	156	300	439	271	61	143	1588
85	57	133	100	54	34	20	60	132	506	237	50	96	1479
91	59	88	165	136	50	49	64	77	308	465	160	121	1742
99	33	66	98	42	67	54	107	75	412	669	353	147	2123
109	47	82	101	57	20	3	11	6	81	508	261	78	1255
113	100	179	183	111	50	21	18	10	64	963	317	50	2066
120	124	264	370	149	50	9	10	10	21	365	143	84	1599
127	203	293	250	236	85	55	27	9	11	109	146	87	1511
134	117	242	219	130	131	77	45	11	10	125	430	238	1775
146	68	173	200	108	57	21	32	38	70	87	268	325	1447
159	42	128	130	71	40	7	9	1	5	17	51	67	568
166	133	203	160	124	99	36	9	11	20	138	145	125	1203
178	72	220	130	119	60	33	26	22	22	36	137	137	1014
191	40	73	52	35	38	57	45	23	19	27	37	379	725
203		2	11	18	21	16	10	13	15	16	19	193	334
218						1	3	5	5	6	11	219	250
228							3	3	10	6		162	184

and what they describe as 'occasional poor samples'. This is the type of variation associated with sample days that Braner and Hairston (1989) allow for by assuming a gamma distribution for daily sampling intensities. Rigler and Cooley adjusted for poor samples by multiplying some sample counts by correction factors based on the trend in the total population count.

Example 2.6   Sampling Four Grasshopper Populations on a Common Site

Another example of the collection of stage-frequency data is described by Bradley (1985). His interest was in the comparative demography of four species of grasshopper, *Chorthippus parallelus*, *C. brunneus*, *Omocestus viridulus* and *O. lineatus*, on Selsey Common, in the Cotswold Hills of Britain.

The study area was defined by setting out a transect line of length 400 m at an approximately constant altitude, and then setting up parallel lines 100 m above and 100 m below this line, to give a sampling area of approximately 80 000 m^2. To sample grasshoppers, quadrats were located at random at each sample time. The quadrats consisted of a 1-m^2 frame covered with a funnel of butter muslin. Care was taken to ensure that no grasshoppers escaped when the frame was set down. An analysis of quadrat counts showed that the grasshoppers were not randomly distributed, although the pattern of distribution was broadly similar for all species.

Bradley gave some consideration to the idea of estimating grasshopper numbers using mark–recapture methods (see Section 2.13). However, this sampling method had to be rejected because marks are lost through moulting as well as death.

2.12   SAMPLING SPECIES ASSEMBLAGES

Some of the most difficult sampling problems occur when relative numbers need to be estimated for several species in an area, an example being the determination of the abundance of all arthropods in a forest canopy. Different species of arthropods have different behaviour and any one of the standard sampling methods will probably not be satisfactory for all of them (Morrison *et al.*, 1989). Because many sampling methods (e.g., using sticky traps) do not provide numbers that can be related to leaf areas, sampling using several of these methods produces results that are difficult to combine. The following example suggests one way of tackling the problem.

Example 2.7   Sampling Arthropods in a Forest

Cooper (1989) stratified a forest by tree species, randomly sampled trees from each of the strata, and carefully counted by sight the arthropods of different species in a patch on each sampled tree. The patches were at heights ranging from 3 to 12 m, the assumption being made that the distribution of arthropods was similar at these low levels to what would be found in upper canopies. The ultimate aim was to express numbers in terms of densities per square metre of leaf area. To this end, the number of leaves in each patch and also the lengths of the leaves were recorded. Known regression relationships between leaf lengths and areas (different for each tree species) were then used to estimate the total leaf area in each tree patch. In that way, the average density per square metre could be estimated for different arthropod taxa, for each tree species (stratum). Total forest densities were then estimated by weighting the results for each tree species by the basal area of that species in the study area, on the assumption that the basal area was proportional to the crown volume.

This example illustrates a number of the ideas and methods discussed in this chapter. The target population (all arthropods in the forest) differs from the sampled population (arthropods in the easily counted patches in trees) but it was assumed that densities per square metre are about the same for both populations. The sampling unit is a patch (cluster) of leaves. The total area of each patch is used as the measure of cluster size. Stratification was by tree species with the area covered by different species replacing the number of trees in different strata for weighting purposes.

2.13   SPECIAL SAMPLING METHODS

The methods described above based on the theory of sampling finite populations are appropriate when that population consists of individuals in known, fixed locations. Random sampling is then possible from a sampling frame or a map. However, with many biological populations the position of individuals is not known in advance and, with mobile animals, changes with time. This has led biologists to devise a number of specialized sampling methods for population estimation. Three of the most important of these will be briefly described in this section.

A characteristic of these methods is that individuals are sampled as they are encountered, and it is impossible to know in advance how many will be found. In some cases, the encounters occur according to a fixed sampling plan whereas at other times it is literally a case of collecting what can be found. Thus, in line–transect sampling of a grasshopper population an observer walks along a predetermined line and records the number of grasshoppers flushed. On the other hand, in mark–recapture sampling of the dragonflies around a pond an attempt may be made to net every individual seen in the area, with no particular set path being traversed.

Removal trapping is one way for estimating the size of a mobile animal population in cases where only a fraction of the population is liable to be captured on any particular sampling occasion. If a series of samples is taken and captured individuals are removed from the population then there will be fewer and fewer individuals available for capture, which should be reflected in decreasing sample sizes. In principle, the trend in numbers captured can be used to estimate the total population size. 'Removal' does not necessarily entail physical removal from the population. It is sufficient for animals to be marked on first capture so that they can be ignored when they are captured again.

Fairly strong assumptions have to be made for the analysis of removal data using one of the standard methods. These are (1) that the population size remains constant during the trapping period, or, at least, natality, mortality and migration are negligible and (2) that the probability of an animal being captured in a sample is the same for all uncaptured animals on all sampling occasions. Given these assumptions, maximum likelihood estimates of the total population size and the capture probability can be determined readily (Southwood, 1978, p. 230; Seber, 1982, p. 309).

With the simplest type of **mark–recapture sampling** a sample is taken from a population, the individuals are marked and returned to the population, and a second sample is taken. Under certain circumstances it is then possible to estimate the total number of individuals in the population. The argument used is that if n_1 individuals are in the first sample, n_2 individuals are in the second sample, and m marked individuals are found in the second sample, then m/n_2 is an estimate of the proportion of marked animals in the population. Hence, if N is the population size, $m/n_2 \simeq n_1/N$, so that $N \simeq n_1 n_2 / m$. Implicit assumptions are that the second sample is random, and there are no deaths or births in the population.

More complicated schemes involve a series of samples being taken with individuals being marked when they are first caught so that all their recaptures can be recorded. In principle, it is then possible to monitor a changing population size and estimate birth and death rates. This does, however, require a large proportion of the population to be captured in each sample if the population size is small. A requirement is that marks remain on animals until they die or emigrate from the study region. This means that in general mark–recapture methods are not useful with populations where individuals will lose their marks on moulting (see Example 2.6).

One of the main areas of application of mark–recapture methods has been for the estimation of parameters of populations of flying insects. An early example is Fisher and Ford's (1947) study of the moth *Panaxia dominula* at Cothill in Berkshire, UK, where the population size had to be estimated each year in order to establish whether the observed fluctuations in the proportion of a mutant gene can be explained by random genetic drift. Even earlier, Jackson (1939) had used these methods to study the dynamics of tsetse fly populations.

Several books are devoted solely or almost entirely to the design and analysis of mark–recapture studies (Begon, 1979; Blower *et al.*, 1981; Seber, 1982). Broadly, the area can be divided into methods applicable to closed populations (i.e., those with negligible natality, death and migration) and those applicable to open populations (i.e., those with new entries, deaths and permanent migration). In the former case, models are available that allow capture probabilities to vary with time or individual or capture history, and some combinations of these (Otis *et al.*, 1978). In the latter case, the Jolly–Seber model (Jolly, 1965; Seber, 1965) has had a prominent place in theory. This requires rather good data and some more recent papers have considered how estimation can be improved by placing restrictions on some parameters (e.g., Jolly, 1979, 1982; Crosbie and Manly, 1985; Brownie *et al.*, 1986). Cormack (1981) and Burnham (1989) have discussed the estimation and comparison of models using standard computer packages (GLIM and SAS, respectively). Arnason and Baniuk (1980) and Clobert *et al.* (1987) describe computer packages designed specifically for manipulating and analysing mark–recapture data.

With **transect sampling**, methods can be classified into three categories (Eberhardt, 1978). With **line–intercept sampling**, what is recorded is the number of objects that are crossed by a line through the sampling region. This can be used to sample large immobile objects that are readily seen. With **strip–transect sampling**, objects are counted in a strip running through the sampling region. This is appropriate with smaller objects that can reasonably be considered to be located at dimensionless points. With **line–transect sampling** an observer moves in a straight line through the area covered by a population and counts how many members of the population are seen. In some cases, animals will move when they are disturbed. In other cases, it is just a case of counting those that are sighted whilst stationary. Generally, the probability of an individual being counted will depend on its distance from the transect line and the data have to be used to estimate the probability function involved in order that the observed count can be related to numbers per unit area. Reviews of these transect sampling schemes are provided by DeVries (1979), Eberhardt (1978), Gates (1979), Seber (1982, p. 460) and Southwood (1978, p. 237).

EXERCISES

1. Divide the population in Table 2.2 into two strata of 40 rows and 20 columns. Compare the accuracy of a stratified sample with 10 taken from each stratum and a simple random sample of 20.

2. The model population in Table 2.3 can be regarded as being laid out in an array of 40 rows and 6 columns. Divide the population into four strata in an appropriate manner and take a simple random sample of five units from each of these. Estimate the Y total using a combined ratio estimator. See how the accuracy compares with that of the simple random sample in Example 2.3.

3 Maximum likelihood estimation of models

3.1 THE METHOD OF MAXIMUM LIKELIHOOD

The method of maximum likelihood (ML) is a standard way of estimating parameters of statistical models. It involves choosing as estimates the parameter values that make the probability of obtaining the observed data (the **likelihood function**) as large as possible. There are several reasons why this method is favoured by many statisticians. Mainly, it is because under fairly general conditions, with large samples, ML estimators are unbiased, have the smallest possible variance, and have variances and covariances that can be approximated fairly easily. Also, the method provides a systematic way of determining estimates that can be applied purely numerically if necessary. A disadvantage in some cases is that estimates can only be determined after lengthy iterative calculations that may not converge on stable values.

A specialist text such as that of Dobson (1983) should be consulted for a comprehensive introduction to the theory of ML. The treatment in this chapter concentrates on those aspects that are most important for fitting models to data, on the assumption that this will be done using a standard computer program. Furthermore, because this book is concerned almost entirely with count data, the discussion will be restricted to these cases. The main application, which is considered in the following two chapters, is to modelling stage-frequency data.

3.2 MODELS FOR COUNT DATA

The likelihood function is the probability of observing the data, regarded as a function of any unknown parameters. For example, suppose that an insect population is distributed randomly over an area that is divided into a large number of 1-m^2 quadrats. The counts in the quadrats are then expected to have a Poisson distribution with mean μ. If a random sample of n quadrats is selected, and gives insect counts of $r_1, r_2, \ldots, r_n$, then the probability of observing the ith count is

$$P(r_i) = \exp(-\mu)\mu^{r_i}/r_i!,$$

and the probability of observing the whole sample is the product of probabilities

$$L(\mu) = \prod_{i=1}^{n} P(r_i) = \prod_{i=1}^{n} \exp(-\mu)\mu^{r_i}/r_i!.$$

This is a function of the Poisson mean μ. If this mean is unknown then the ML estimate $\hat{\mu}$ of μ is the value that makes $L(\mu)$ as large as possible. It can be shown that this value is the mean $\Sigma r_i/n$ of the observed counts.

In this example it is possible to determine the ML estimate by calculus because it is the value of μ for which the first derivative of $L(\mu)$ is zero. However, often, and particularly when the likelihood function depends on several unknown parameters, no such explicit maximum can be determined. Instead, the likelihood function must be maximized by an iterative numerical method (Manly, 1985a, p. 405). As noted before, there is no guarantee that the iterations will converge in all cases.

There are three models for count data that are particularly useful with stage-frequency data. These are the Poisson model, the multinomial model, and the product multinomial model. As will now be shown, estimation problems can often be looked on as being essentially the same, irrespective of which of these three models seems to be most appropriate.

The **Poisson model** is a generalization of the one just considered, where the data are the counts $r_1, r_2, \ldots, r_K$ in K classes. These counts might, for example, be for K development stages, for K different species, for males and females (with $K = 2$), etc. The assumption is made that r_i is a random value from a Poisson distribution with mean μ_i and that the distribution is independent of the other counts. This model can be derived theoretically on the assumption that there are a large number of individuals in class i that could be included in the count r_i, with each of these having the same small probability of inclusion.

In modelling situations the mean μ_i can be written as $\mu_i(\theta_1, \theta_2, \ldots, \theta_P)$ since it will be a function of unknown parameters $\theta_1, \theta_2, \ldots, \theta_P$ to be estimated. The likelihood function is then the product of the Poisson probabilities of obtaining the counts, which is

$$L = \sum_{i=1}^{K} \exp(-\mu_i)\mu_i^{r_i}.$$

With the **multinomial model** the total count in all classes, $n = r_1 + r_2 + \ldots + r_K$ is regarded as fixed. It is assumed that each of the n sampled individuals has a probability P_i of being in the ith class, independent of all other individuals, with $P_1 + P_2 + \ldots + P_K = 1$. In this case P_i will be a known function of the parameters $\theta_1, \theta_2, \ldots, \theta_P$ to be estimated and the likelihood function to be maximized is the multinomial probability

$$L = N!/(r_1!r_2!\ldots r_K!)P_1^{r_1}P_2^{r_2}\ldots P_K^{r_K}.$$

The relationship between the Poisson and multinomial models is as follows. Suppose that the Poisson model is correct. Then the distribution of the sample counts, conditional on the total count being equal to n, is multinomial with the class probabilities being

$$P_1 = \mu_1 / \sum_{i=1}^{K} \mu_i, \; P_2 = \mu_2 / \sum_{i=1}^{K} \mu_i, \ldots, P_K = \mu_K / \sum_{i=1}^{K} \mu_i,$$

(Manly, 1985a, p. 403). This means that if the Poisson model is thought to be correct then the multinomial distribution can be used to analyse data, with the analysis being conditional on the observed total count.

Table 3.1   A two-way array of counts with individuals in s samples classified into K classes

		Class			
Sample	*1*	*2*	. . .	*K*	*Total*
1	n_{11}	n_{12}	. . .	n_{1K}	n_1
2	n_{21}	n_{22}	. . .	n_{2K}	n_2
.					
.					
.					
s	n_{s1}	n_{s2}	. . .	n_{sK}	n_s

A situation that occurs rather frequently is that there is a two-way array of sample frequencies, as shown in Table 3.1. There are s samples taken at different times and/or places, and the individuals in each sample are placed into K classes according to such characteristics as phenotype, species or developmental stage. The counts n_{ij} might then reasonably be regarded as independent Poisson variables so that the Poisson model applies with the cell means μ_{ij} being functions of unknown parameters $\theta_1, \theta_2, \ldots, \theta_P$. An analysis conditional on the total count n is then possible using a multinomial model for which the probability of an observation in row i and column j is

$$P_{ij}(\theta_1, \theta_2, \ldots, \theta_P) = \mu_{ij} / \sum_{u=1}^{s} \sum_{v=1}^{K} \mu_{uv}. \tag{3.1}$$

Another possible model for data of the form of Table 3.1 is one where the row totals are regarded as being fixed and the distribution of counts within rows have (possibly different) multinomial distributions that depend on parameters $\theta_1, \theta_2, \ldots, \theta_P$. This is then the **product multinomial model** because the probability of obtaining all the counts n_{ij} at the same time is the product of s multinomial probabilities. In this case it is possible to show that the ML estimates of the θ parameters are the same as those obtained from the single multinomial model of equation (3.1) if μ_{ij} is taken as being equal to n_i multiplied by the probability of being in class j in sample i (Manly, 1985a, p. 435).

With field sampling it is unusual to be able to fix sample sizes in advance. Hence, a Poisson model may be more appropriate for describing the sampling process than a multinomial model. However, for estimation purposes it will often be convenient to use the multinomial model with an analysis being regarded as conditional on the total sample count. Indeed, in all the following examples of maximum likelihood estimation it is the multinomial model of this type that has been assumed.

3.3   COMPUTER PROGRAMS

Various computer programs are available for fitting models to count data using the method of maximum likelihood. Some of these, such as GLIM (Nelder, 1975; McCullagh and Nelder, 1983; Manly, 1985a, p. 415), the SAS procedures FREQ and CATMOD (SAS Institute, 1985) and the BMDP procedure 4F (Dixon, 1985) are intended mainly for fitting certain particular types of model. Others such as MAXLIK (Reed and Schull, 1968; Reed, 1969; Manly, 1985a, p. 433), the SAS procedure NLIN (SAS Institute, 1985), and the BMDP procedures P3R and PAR (Dixon, 1985) can be used for a wider class of models.

It is the MAXLIK algorithm that has been used for the maximum likelihood calculations in the chapters that follow. The references just given for this program should be consulted for more details about its use. Here it suffices to say that to use the program it is necessary to provide a FORTRAN subroutine to calculate the expected relative frequencies in K data classes as functions of the unknown parameters to be estimated, together with initial approximations for the ML estimates of these parameters. The program then endeavours to improve these estimates by maximizing the likelihood with an iterative process. Output includes the final estimates, the maximized log-likelihood, approximate standard errors, and approximate correlations between estimators. A multinomial likelihood function is assumed. Programs tailored for different applications are included in a package described in the Preface.

It is usual for programs for maximum likelihood estimation to work in terms of the logarithm of the likelihood function rather than the likelihood function itself because for large sets of data the value of the likelihood function may be extremely small even when it is maximized. Also, as noted below, it is the log-likelihood that is important for comparing models and determining the goodness of fit of models. Of course, the same parameter values maximize both the likelihood and log-likelihood functions.

3.4 MEASURING GOODNESS OF FIT

The goodness of fit of a model fitted by maximum likelihood can be measured by the **deviance**, where this is minus twice the maximized log-likelihood. Maximizing the log-likelihood is equivalent to minimizing the deviance. In using deviances the likelihood function has to be scaled so that $D = 0$ for a model where the observed and expected sample frequencies agree exactly. Small values for deviances are desirable and it is possible to test to see whether a deviance is significantly large by comparing it with critical values for the chi-squared distribution with the appropriate degrees of freedom.

Conventionally, the degrees of freedom are determined by the formula $N - 1 - p$, where N is the total number of frequencies in the data and p is the number of estimated parameters. However, with most stage-frequency data this rather obviously gives too many degrees of freedom because of the large number of counts that cannot really be anything but zero. Consider, for example, Qasrawi's (1966) data obtained from sampling a population of grasshoppers (see Table 1.2). Here there are 29 samples and five stages, and hence 145 stage-frequencies altogether. The formula $N - 1 - p$ therefore suggests that the deviance for a model fitted to these data should have $144 - p$ degrees of freedom. This is clearly too high. For instance, each sample after 11 August increases the degrees of freedom by five even though any reasonable model for the data must give almost zero probabilities for individuals being in stages 1–4 for these samples. Also, stages 2–5 in sample 1 contribute 4 degrees of freedom to the total although again any reasonable model will give these stages very small probabilities.

A more appropriate formula for the degrees of freedom is $R - 1 - p$, where R is the number of stage-frequencies that are potentially non-zero, and p is the number of estimated parameters. For Qasrawi's data, it seems reasonable to include in R the 11 stage 1 counts from 20 May to 2 July, the 13 stage 2 counts from 3 June to 24 July, and so on for the other stages. In other words, for any stage the contribution to R consists of the number of samples for the time period when the stage was present. On this basis the degrees of freedom for Qasrawi's data are $68 - p$, which is only about half the number given by the formula $N - 1 - p$. This approach to determining the degrees of freedom is the one that is used in the examples in the following chapters, with two minor adjustments that are needed only with some sets of data. The first adjustment is applied if there are any cases where a sample includes stage $j + 1$ individuals before any stage j individuals are seen. This suggests that the range of sample times for stage j should begin when stage $j + 1$ is first seen since an individual must pass through stage j to reach stage $j + 1$. The second adjustment is applied if there are any cases where individuals are still in stage $j - 1$ after the last individuals are seen in stage j. This suggests that the range of sample times for stage j should be extended until all individuals have left stage $j - 1$.

It can be argued that, if anything, the method proposed here for determining the degrees of freedom may result in too few degrees of freedom being used. This follows since there may be cases where a stage could potentially have been seen in a sample before the first observed sighting or after the last observed sighting. Nevertheless, the proposed rule for determining the degrees of freedom should give about the correct result for most sets of data and seems to be better than most of the obvious alternatives.

Another approach to measuring goodness of fit involves calculating a Pearson chi-squared statistic, $\Sigma(O - E)^2/E$, where O denotes observed and E denotes expected frequencies. As with the deviance, the degrees of freedom are usually taken as $N - 1 - p$, where N is the total number of frequencies and p is the number of estimated parameters. It is usual to pool cells with small expected frequencies for this type of calculation but an alternative possibility with stage-frequency data involves considering only the frequencies for a stage for the samples when that stage is present, in the same way that has been suggested for deviances. The deviance and Pearson chi-squared then have the same degrees of freedom and are directly comparable. In many cases they will have similar numerical values.

3.5 COMPARING MODELS

One of the advantages of using maximum likelihood to fit models is that it is relatively easy to compare two alternative models in cases where one model is a more general version of the other, with one or more extra parameters. This is done by comparing the **deviances**. Thus suppose that model 1 is the simpler model with a deviance of D_1. Model 2 then has the same parameters as model 1, plus at least one other parameter, and a deviance of D_2. It follows that $D_1 \geq D_2$ since the extra parameters will allow model 2 to be at least as good a fit as model 1. The improvement in fit of model 2 over model 1 can then be measured by $d = D_1 - D_2$, which must be positive or zero. The degrees of freedom for d are the number of extra parameters in model 2 compared to model 1. A significantly large value of d in comparison with the chi-squared distribution indicates that model 2 is an improvement over model 1. An insignificant value of d indicates that the extra parameters in model 2 are unnecessary.

The tests based on the chi-squared distribution that have been described in this and the previous section are based on the asymptotic distributions of deviances and Pearson chi-squared statistics for cases where data counts are 'large'. The precise requirements are difficult to quantify and require further study. All that can be said at present is that they should be satisfied for sets of data with most non-zero observed frequencies being five or more. Unfortunately, this condition is not often met with stage-frequency data. In cases of doubt it is always possible to generate artificial data following the model being fitted and determine the distributions that this model gives for sample statistics. Observed statistics for the real data can then be assessed in comparison with the empirical distributions.

3.6   THE HETEROGENEITY FACTOR

A problem that can be expected to occur frequently with field data is an apparent lack of fit of the model being considered. This is not necessarily an indication that the model is inappropriate. It may be that the assumption of Poisson sampling errors is wrong so that there is extraneous variation in the sample counts, as discussed in Section 2.11 for stage-frequency data. A moment's reflection will show that this can occur very easily. For example, if the individuals in a population have a clustered geographical distribution then there will be a tendency for samples to contain either few or many individuals. Also, in sampling a population over a period of time every effort may be made to maintain a constant sampling fraction but the numbers caught on different days may still depend on the weather to some extent.

In some cases an allowance for extraneous variation can be made by using the **heterogeneity factor**, also mentioned in Section 2.11. The assumption that has to be made in doing this is that the variances of all sample counts are larger than the Poisson variance by a constant multiple. With models estimated by maximum likelihood this multiple can then be estimated by $H = D/(R - 1 - p)$, the deviance divided by its degrees of freedom. An alternative estimate is given by the Pearson chi-squared statistic divided by the degrees of freedom.

Suppose that a population is sampled randomly, so that sampling errors do indeed follow Poisson distributions, but all frequencies are then multiplied by $C > 1$. If the correct model is then fitted to the data it can be expected to display extraneous variance, and the expected value of H is C. This interpretation of H indicates that when extraneous variance appears to be present in data it can be corrected for by dividing the sample sizes used to collect the data by H, to get effective sample sizes. Since standard errors are generally proportional to reciprocals of square roots of sample sizes, they should be multiplied by $\sqrt{H}$ to be adjusted for extraneous variance. Variances and covariances of estimators should be multiplied by H. Correlations should not require any adjustment.

Comparing two models fitted by maximum likelihood is complicated by the presence of extraneous variation. Suppose model 1 is a simple model with p_1 parameters and model 2 is a more general model with p_2 parameters, these being the same as those of model 1 plus some others. Also, suppose that D_1 is the deviance of model 1 and D_2 is the deviance of model 2, with D_2, being significantly large compared with the chi-squared distribution with $R - 1 - p_2$ degrees of freedom. Although the deviance of model 2 is significantly large, it may nevertheless be reasonable to assume that model 2 is correct, with extraneous sampling variance being present. An important question then may be whether the simpler model 1 is also correct. This can then be tested by setting up an analysis-of-variance table, as shown below.

Source of variation	Deviance	Degrees of freedom	Mean deviance
Extra parameters in model 2	$D_1 - D_2$	$p_2 - p_1$	$(D_1 - D_2)/(p_2 - p_1)$
Error (fit of model 2)	D_2	$R - 1 - p_2$	$D_2/(R - 1 - p_2)$
Total	D_1	$R - 1 - p_1$	

The significance of the extra parameters in model 2 can be assessed by comparing the ratio of mean deviances

$$F = \{(D_1 - D_2)/(p_2 - p_1)\}/\{D_2/(R - 1 - p_2)\}$$

to the F distribution with $p_2 - p_1$ and $R - 1 - p_2$ degrees of freedom. A significantly large value indicates that the extra parameters of model 1 are needed to describe the data.

The use of the heterogeneity factor can be extended beyond models for count data with extraneous variance. In any case, where the variances of observations are believed to be proportional to their expected values, data can be analysed as if they are counts from Poisson distributions with variances equal to expected values. The heterogeneity factor gives an estimate of the multiplication factor for variances and can be used to adjust standard errors and in the comparison of models, as indicated above. The literature on this type of adjustment is reviewed by Burnham *et al.* (1987); see also McCullagh and Nelder (1983, p. 131).

Of course, using a heterogeneity factor is not a panacea for all cases where data do not fit a model. Often, the model is simply unrealistic. It is then completely misleading to hide the poor fit by introducing the factor. For example, the data on *Skistodiaptomus oregonensis* shown in Table 2.5 appear to exhibit more variation than is expected from Poisson sampling errors. However, this does not seem to be a case of the excess variation being independent for all the stage-frequencies. Rather, the counts for all the stages seem to tend either to all be high or all be low for some of the samples. Using a heterogeneity factor will not account for correlations between sampling errors and hence will give only a rather approximate adjustment for extraneous variance. It would be better to model the extraneous variance explicitly, possibly using Braner and Hairston's (1989) approach.

4 Analysis of multi-cohort stage-frequency data

4.1 MULTI-COHORT STAGE-FREQUENCY DATA

This chapter is concerned with the analysis of stage-frequency data. As has been discussed in the previous chapters, such data are of counts or estimates of the numbers of individuals in the various development stages in a population, or in a fraction of a population, at a series of points in time. Interest usually centres on obtaining estimates of some or all of the following parameters:

1. The total number of individuals entering each stage.
2. The average time spent in each stage, and possibly the distributions of stage durations.
3. The probabilities of surviving each stage (the stage-specific survival rates).
4. The mean time of entry to each stage.
5. Unit time survival rates.

A distinction can be made between multi-cohort and single cohort stage-frequency data. With multi-cohort data, individuals are entering the population for a substantial part of the sampling time. This means that the entry time distribution (which will generally be unknown) is confounded with the distributions of the durations of stages to some extent. This can be expected to make the estimation of the distributions of stage durations somewhat difficult. Deaths are also offset by new entries, which makes survival rates difficult to estimate unless data are available for some time after all entries have been made. On the other hand, with single cohort data all individuals enter the population at the start of the sampling period. In principle, this should make estimation much easier. Multi-cohort data are considered in the present chapter while the following chapter deals with single cohort data.

An example of multi-cohort data is shown in Table 1.2. These are the stage-frequency counts obtained by Qasrawi (1966) when he sampled a population of the grasshopper *Chorthippus parallelus* on East Budleigh Common, Devonshire, in 1964. *Chorthippus parallelus* hatches from eggs and then passes through four instars before reaching the adult stage. Because Qasrawi sampled a fraction 0.002 of his defined site (except as noted in the table), population frequencies can be estimated as the sample frequencies multiplied by $1/0.002 = 500$. However, for most purposes the stage-frequencies can be used as they stand.

Simply inspecting the frequencies is quite instructive. It can be seen that when sampling began on 20 May all the grasshoppers found were in instar 1. Subsequently, the numbers in this stage increased to a peak on 29 May and then declined to zero from 9 July onwards. Grasshoppers were first seen in instar 2 on 3 June. Numbers in this stage peaked on 25 June, and declined down to zero. The other two instar stages also show this pattern of increasing numbers followed by a decrease to zero. When sampling ceased on 23 September the only grasshopper seen was in the adult stage. The sample frequencies are quite small and very likely contain substantial sampling errors. Nevertheless, there is a clear picture of grasshoppers beginning to hatch just before the 20 May sample and continuing to hatch until about the middle of June. After hatching, the grasshoppers proceeded to pass gradually through the instar stages to the adult stage. The individuals that hatched first reached the adult stage early in July while the late-hatching individuals reached the adult stage by about the middle of August.

Ashford *et al.* (1970) argued that migration across the boundaries of the study area can be neglected when considering these results because of the limited mobility of the grasshoppers and the similarity of the surrounding area. On that assumption, the generally decreasing total sample frequency after entries to stage 1 ended can be assumed to be a reflection of the decreasing number of survivors in the population.

Another example is shown in Table 4.1. This is van Straalen's (1982) data on the collembolan *Orchesella cincta*. Here the 'stages' are size classes, rather than different parts of the life cycle. For example, stage 1 consists of individuals with a length of between 0.73 and 0.90 mm. In principle, there is no reason why stages should not be defined in this way providing that individuals that live long enough will eventually reach the final stage.

As has been discussed in Section 2.11, there are many complications that can occur with stage-frequency data, particularly for data collected from wild populations. It may be impossible to sample all stages equally well. The adults, in particular, may be more mobile than the individuals in the other stages. Adult numbers may then have to be determined by quite a different sampling method to that used with other stages (e.g., mark–recapture sampling instead of sweep net sampling). In that case it may well turn out that the estimated adult numbers cannot be reconciled with the estimated numbers in the other stages. The adults may even disperse so that they cannot be counted at all, as was the case for a study of the mosquito *Aedes cantans* described by Lakhani and Service (1974).

Many models for stage-frequency data assume that the daily survival rate

Table 4.1 Data for a generation of the collembolan *Orchesella cincta* in a forest near Dronten, The Netherlands (from van Straalen, 1982). The 'stages' are seven size classes with the lower limits 0.73, 0.90, 1.35, 1.80, 2.25, 2.70 and 3.15 mm. The t' values are physiological times

		Stage							
Week	t'	1	2	3	4	5	6	7	Total
1	4.6	34							34
3	6.8	588	124	47	1				760
5	8.5	378	199	55	33	1	2		668
7	10.6	175	302	193	73	11	2		756
9	12.5	15	89	140	86	42	7		379
11	14.8	25	53	85	137	117	42	1	460
13	16.8		12	24	51	89	65	3	244
15	18.7			2	14	44	25	0	85
17	20.8				27	31	34	6	98
19	22.2				18	34	52	13	117
21	23.9				7	2	5	1	15
25	25.2				4	2	3	1	10
29	26.2						1	1	2

is constant either for all individuals throughout the sampling period, or for all individuals in the same developmental stage. Although these assumptions are often reasonable, situations do arise where they are clearly untenable. For example, if an insect population is sprayed with an insecticide during a sampling programme it is almost certain that one of the effects will be a low survival rate from the time of spraying until the time that the insecticide has dispersed. Birley (1977) gives an example of this type of 'catastrophe' with a population of the froghopper *Aeneolamia varia saccharina*.

4.2 TEMPERATURE EFFECTS

One reason why daily survival rates and other population parameters may not be constant is that they are linked to environmental conditions, particularly ambient temperatures. A common situation is that physiological time passes slowly at low temperatures and quickly at high temperatures. It may then be appropriate to replace 'ordinary' time with 'physiological' time before starting to analyse data (Pajunen, 1983). The effects of doing this can be quite large. For example, in his study of *Orchesella cincta* van Straalen (1985) concluded that the months November to March in 1979 corresponded to only 3 weeks of physiological time.

Of course, there is no reason why the physiological time for development of individuals through stages should correspond to the effective time as far as the survival of those individuals is concerned. Survival rates may be more constant in terms of calendar time than in terms of physiological time. If that occurs then an appropriate model for data will need to allow for both calendar and physiological timescales.

Physiological times are usually measured in 'degree days', the idea being that the amount of development accumulated by a population is proportional to the amount of time that ambient temperatures have exceeded a certain threshold, below which no development occurs. This model for temperature effects is undoubtedly useful, but is known to have its limitations. Typically, development occurs only within a definite temperature range. Development stops when the temperature is at or below the lower limit, increases approximately linearly with temperature as the temperature increases from the lower limit, through moderate temperatures, and then begins to slow down as an upper limit to temperature is reached (Logan *et al.*, 1976; Curry *et al.*, 1978; Wagner *et al.*, 1984a). Mortality rates become very high as the temperature reaches the maximum limit, making developmental rates difficult to determine. See also Benton (1988) and Cherrill and Begon (1989).

Generally, the individuals in a population will not all develop at the same rate. Several authors have discussed the effects of temperature on the distribution of developmental rates within a population. Distributions of times spent in stages are generally skewed to the right, possibly because the developmental rate at a constant temperature is normally distributed (Sharpe *et al.*, 1977). Assuming that the same form of distribution holds at different temperatures, modelling temperature effects involves choosing this distribution and deciding how the parameters change with temperature. A useful simplification occurs if it can be assumed that the distributions are the same shape (Wagner *et al.*, 1984b, 1985).

Needless to say, temperature and development relationships estimated from laboratory experiments at a constant temperature may not apply in the field. For this reason it may be best to analyse field data using degree days for physiological time, without relying on laboratory results at all. This should work reasonably well, providing that the temperature fluctuations experienced in the field are not too extreme during the study period.

4.3 EFFECT OF MORTALITY ON STAGE DURATIONS

Braner (1988) and Braner and Hairston (1989) have noted the distinction between stage durations that are observed in populations subjected to mortality, and the durations that would be observed in the absence of mortality. These will not generally be the same if the duration of stages is not the same for all individuals. Clearly, if mortality occurs at a more or less constant rate throughout the time that individuals are in a stage, then individuals that only spend a short time in the stage are more likely to survive the stage than are individuals that spend a long time in the stage. The survivors from a stage therefore represent a biased sample for the estimation of the mortality-free distribution of durations.

Unless the mortality rate in a stage is high, the difference between the distribution of stage durations for survivors and the mortality-free distribution should be relatively small in comparison with typical errors of estimation. Nevertheless, the distinction between the distributions is an important one. It can be noted that most of the estimates related to stage durations that are described below are for the durations that would be observed in the absence of mortality.

4.4   METHODS FOR ANALYSING MULTI-COHORT STAGE-FREQUENCY DATA

A comprehensive review of the large number of methods that have been proposed for analysing multi-cohort stage-frequency data is beyond the scope of this book. Instead, a few particularly useful methods will be described in some detail. However, it is worthwhile to consider the various approaches that are available, and a list of these is provided in Table 4.2. More details about many of the methods in this list can be found in Manly (1989a); see also Schaalje and van der Vaart (1989) for a classification of some of the models used.

The table includes a brief indication of the assumptions made by the methods cited, what they estimate, and the principle used. Note that some methods require more information than just the observed sample stage-frequencies. For example, the Richards *et al.* (1960) method can be used only if the number entering stage 1 and the durations of all stages are known.

4.5   ASSESSING ESTIMATES BY SIMULATION

Stage-frequency data are derived under many different circumstances. As a consequence, it is difficult to discuss the determination of the accuracy of estimates and the testing of assumptions without referring to particular examples. However, it is appropriate to say a few words here about simulation, which often provides the simplest and most versatile method for addressing these questions.

If the mechanism generating data is known then many artificial sets of data can be produced on a computer with parameter values equal to those estimated from a particular set of data. These artificial sets can be analysed in order to generate approximate sampling distributions for estimates of population parameters. In this way, biases and standard errors can be determined. Also, goodness-of-fit statistics such as the sum of squares of deviations between observed and expected stage frequencies can be calculated both for the observed data and the artificial data. If the observed statistic falls well within the distribution for the generated data then the model can be considered to give a reasonable fit to the observed data. If the observed statistic is very large when compared to the generated distribution, then the model is questionable. If the observed statistic is very small when compared to the generated distribution, then this suggests that the model includes too many sources of variation.

Several different approaches to simulation can be used. Obviously, if the data stage-frequencies have been determined by some rather special method such as capture–recapture sampling then the simulation will have to attempt to model this process. However, in many cases the data values are obtained simply by sampling a small fraction of a large population. Assuming random sampling, the sample frequencies will then be independent Poisson variates with mean values equal to the population stage-frequencies multiplied by the sampling fraction (see Section 2.11). In that case, four possible methods of simulation suggest themselves.

Method 1 involves simulating the population individual by individual, and then giving each live individual a probability of being 'captured' at each sample time. This has the advantages of allowing any desired distributions for the time of entry to stage 1 and the durations of stages, and making it possible to study the variation in estimates caused by stochastic effects in the population being sampled as well as variation caused by sampling errors. If stage-frequency data are obtained by sampling an appreciable fraction of a population, so that these stochastic effects are liable to be important, then a method 1 simulation is the only realistic way to determine the properties of estimation procedures. The main disadvantage of this method of simulation is that generating samples from large populations takes a long time. This was the approach used by Manly (1974b) for the comparison of several alternative methods of estimation.

Method 2 is attractive for simulating samples from a large population. It involves using a specific model to calculate expected population stage-frequencies, and then determining sample stage-frequencies by generating independent Poisson variates with these expected values. Because the expected frequencies are determined as functions of the entry distribution to stage 1 and the distributions of stage durations, this method assumes that the population being sampled is large enough to make stochastic variation in population values negligible.

Method 3 is very simple. It involves assuming that the duration of stages is the same for all individuals. The individuals that enter between two sample

Table 4.2 Summary of methods that have been proposed for analysing stage-frequency data

References	Parameters estimated	Comments
Richards and Waloff (1954)	Numbers entering stages; unit time survival rate	Uses linear regression to relate the number in a stage and all higher stages to time. Assumes a constant survival rate in all stages
Richards *et al.* (1960)	Unit time survival rates in stages	The number entering stage 1 and the durations of stages must be known. The area under the stage-frequency curve is equated to its expected value
Dempster (1961)	Total number entering stage 1; unit time survival rates in stages	Changes in the total stage-frequencies are related to proportions entering stage 1 between samples and survival rates by a linear regression. Entry rates to stage 1 must be known
Southwood and Jepson (1962); Sawyer and Haynes (1984)	Numbers entering stages	The mean durations of stages must be known. Based on an equation that holds only if all mortality occurs at stage transitions
Kiritani and Nakasuji (1967); Manly (1976, 1977a, 1985a)	Numbers entering stages; unit time survival rate; durations of stages	Relates the area under the stage-frequency curve and the time–stage-frequency curve to population parameters. Assumes a constant survival rate in all stages
Kobayashi (1968)	Numbers entering stages	Deaths are apportioned to different stages by a two-part correction process. The number entering stage 1 must be known

Table 4.2 – contd.

References	Parameters estimated	Comments
Read and Ashford (1968); Ashford *et al.* (1970)	Numbers entering stages; unit time survival rate; distributions of stage durations	The first model with a proper statistical model. Estimation by the method of maximum likelihood
Rigler and Cooley (1974)	Numbers entering stages; durations of stages	There has been some discussion about the validity of this method (Hairston and Twombly, 1985; Aksnes and Hoisaeter, 1987; Hairston *et al.*, 1987; Saunders and Lewis, 1987)
Lakhani and Service (1974)	Survival rates in different stages	Equations relating areas under stage-frequency curves to survival rates are solved. Durations of stages and the proportion surviving to the last stage must be known
Manly (1974a)	Numbers entering stages; unit time survival rate; durations of stages	Uses a non-linear regression to relate the number in a stage and all higher stages to population parameters
Ruesink (1975)	Stage-specific survival rates that vary with time	Durations of stages must be known
Birley (1977); Bellows and Birley (1981)	Survival rates in stages; numbers entering stages	The rate of entry to stage 1 must be known. Estimation by non-linear regression
Derr and Ord (1979)	Stage-specific survival rates that vary with time	Durations of stages must be known
Kempton (1979)	Distributions of durations of stages; time-dependent survival rates; numbers entering stages	Distributions of entry times to stages can take various forms. The survival function is not stage-dependent

Table 4.2 – contd.

References	Parameters estimated	Comments
Mills (1981a, 1981b)	Mean durations of stages; numbers entering stages	One duration, or the ratio of two durations must be known
Bellows *et al.* (1982)	Unit time survival rates that can vary with time; distributions of stage durations; numbers entering stages	Survival rates are estimated by regressing total stage-frequencies against time. The Bellows and Birley (1981) model is then used to estimate other parameters
van Straalen (1982, 1985)	Numbers entering stages; unit time survival rate; durations of stages; a growth function	Assumes that each individual has an associated measurable development variable that increases with time according to the growth function. Temperature effects are allowed for by using physiological time. Estimation by the method of maximum likelihood
Osawa *et al.* (1983); Stedinger *et al.* (1985); Dennis *et al.* (1986); Kemp *et al.* (1986); Dennis and Kemp (1989)	Proportion of the population in different stages at different times (and places)	Estimates a normal or logistic function for the distribution of a development variable at different times. Estimation by the method of maximum likelihood. Temperature effects are allowed for by using physiological time
Shoemaker *et al.* (1986)	Recruitment rates for stage 1; stage-specific survival rates; a unit time survival rate; the durations of stages	A non-linear regression relating stage-frequencies to recruitment and survival is estimated. Temperature effects are allowed for by using physiological time
Manly (1987)	Numbers entering stages; a unit time survival rate in each stage; durations of stages	Uses a linear regression to relate the number of individuals in one sample, to the numbers in different stages in the previous sample

Table 4.2 – contd.

References	Parameters estimated	Comments
Munholland (1988); Kemp *et al.* (1989); Munholland *et al.* (1989)	Proportions of the population in different stages at different times; survival rate that can depend on time	A development of the model of Osawa and others given above that allows for mortality. Estimation is by the method of maximum likelihood. Temperature effects are allowed for by using physiological time
Braner (1988); Braner and Hairston (1989)	Numbers entering stages; stage-specific or a constant survival rate; durations of stages	A gamma model is used to model variation in sampling intensities with time. Normal distributions are assumed for stage durations. Estimation by maximum likelihood
Manly (1989a)	Numbers entering stages; a unit time survival rate; distribution of durations of stages; distribution of a development variable; amount of development needed to enter each stage	Assumes that each individual has an unobservable value for a development variable with this being normally distributed at the time of the first sample. Estimation by maximum likelihood

times can then be allowed to enter at (say) five, equally spaced times between the samples and it is quite easy to calculate their expected contributions to stage-frequencies from then on, taking into account the survival rates in different stages and the durations of stages. Once expected frequencies are determined, sample frequencies can be taken as independent Poisson variates with these means. Like method 2, method 3 simulation is only appropriate when samples consist of a small fraction of a large population.

Method 4 is even simpler. A simulated set of data is generated by replacing each observed stage-frequency with a value from the Poisson distribution with a mean equal to this observed value. This is equivalent to random sampling from a population with expected stage-frequencies equal to the observed ones. The main advantage of this method of simulation is that no

particular model has to be assumed for the population. Nevertheless, the
simulated data should display the amount of variation that is to be expected
from sampling errors, assuming that stochastic variation is not important in
comparison with this.

4.6   THE KIRITANI–NAKASUJI–MANLY (KNM) METHOD OF ANALYSIS

The remainder of this chapter will be devoted to the closer examination of
three particularly useful approaches to the analysis of multi-cohort data.
The first of these is a relatively simple method for cases where three
conditions are satisfied. First, the survival rate per unit time should be the
same in all stages for the entire sampling period. Second, sampling should be
started at or shortly after the time when individuals begin to enter stage 1,
and continues until all or almost all individuals are dead. Third, population
losses through migration should be negligible. These conditions allow the
use of a type of moment estimation involving far less calculation than the
other methods of analysis that will be considered in detail. In fact, the
calculations can be done easily enough on a hand calculator. If the second
condition is not satisfied then an iterative method for applying the KNM
method can be used, but this complicates matters considerably.

When Kiritani and Nakasuji (1967) first described their method for
estimating stage-specific survival rates it had the restriction of requiring
samples to be taken at fixed intervals of time for the entire period that a
population is present, and only provided estimates of numbers entering
stages. Later, the calculations were modified to allow for irregular sampling
times and the estimation of other population parameters (Manly, 1976,
1977a, 1985b).

The model for the Kiritani–Nakasuji–Manly (KNM) method assumes that
the population being sampled is large enough for any changes in that
population (e.g., the daily survival rate) to operate deterministically. This
makes it possible to write down equations that give relationships between
population parameters and population stage-frequencies. The assumption
then made is that these equations hold approximately when sample stage-
frequencies replace population values.

An important equation is

$$f_j(t) = M_j \int_{t-a_j}^{t} g_j(x) \exp\{-\theta(t-x)\} dx \qquad (4.1)$$

where $f_j(t)$ is the number in the population in stage j for the fraction of the
population sampled at time t (i.e., the expected value of the sample stage j
frequency at that time); M_j is the number entering stage j in the sampled
fraction of the population; $g_j(x)$ is the probability density function for the

time of entry to stage j; $\exp(-\theta)$ is the survival rate per unit time; and a_j is the duration of stage j, which is assumed to be the same for all individuals that survive through the stage. What this equation says is that the individuals in stage j at time t consist of the individuals that enter between times $t - a_j$ and t that remain alive. The term $\exp\{-\theta(t - x)\}$ is the probability of survival from an entry time x until time t; $M_j g_j(x) dx$ can be thought of as the number entering in the small time interval x to $x + dx$. Integration is then equivalent to adding up the contribution to $f_j(t)$ from individuals with different entry times.

If both sides of equation (4.1) are integrated for t between plus and minus infinity then the probability density function $g_j(x)$ disappears and the result obtained is

$$A_j = \int_{-(\infty)}^{+(\infty)} f_j(t) \, dt = M_j\{1 - \exp(-\theta a_j)\}/\theta. \tag{4.2}$$

This shows how the area under a stage-frequency curve (A_j) is related to the number entering the stage (M_j), the survival parameter (θ), and the duration of the stage (a_j).

Another useful equation is found by multiplying $f_j(t)$ by t before integrating. This yields

$$D_j = \int_{-(\infty)}^{+(\infty)} t f_j(t) \, dt$$

$$= [\mu_j + 1/\theta - a_j \exp(-\theta a_j)/\{1 - \exp(-\theta a_j)\}] A_j, \tag{4.3}$$

which relates the area under the $t f_j(t)$ curve (D_j) to the mean time of entry to the stage (μ_j), the survival parameter (θ), and A_j.

Equations (4.2) and (4.3) can be applied to accumulated stage-frequency data. Thus, if A_j^* is the area under the stage-frequency curve $F_j(t)$ for stages $j, j+1, \ldots, q$ combined, and D_j^* is the area under corresponding time–stage-frequency curve $t F_j(t)$, then there are the relationships

$$A_j^* = M_j \theta, \tag{4.4}$$

and

$$D_j^* = (\mu_j + 1/\theta) A_j^*. \tag{4.5}$$

These are derived by noting that the combined stages include the final stage, which has an infinite duration since all losses are through death. Putting $a_j = \infty$ in equations (4.2) and (4.3) gives the required results.

From the last two equations it follows that

$$w_j = A_{j+1}^*/A_j^* = M_{j+1}/M_j, \qquad j = 1, 2, \ldots, q-1, \tag{4.6}$$

is the stage-specific survival rate for stage j;

$$\theta = -\log_e(A_q^*/A_1^*)/(B_q^* - B_1^*), \tag{4.7}$$

is the survival parameter;

$$a_j = -\log_e(w_j)/\theta, \qquad j = 1, 2, \ldots, \overset{\cdot}{q} - 1, \qquad (4.8)$$

is the duration of stage j; and

$$M_j = A_j^* \theta, \qquad j = 1, 2, \ldots, q, \qquad (4.9)$$

is the number entering stage j, where $B_j^* = D_j^*/A_j^*$.

The values A_j^* and D_j^* in equations (4.6) to (4.9) can be estimated from data using the trapezoidal rule. This is done most easily when observation times are equally spaced at t_1, $t_2 = t_1 + h$, $t_3 = t_1 + 2h$, ..., $t_n = t_1 + (n-1)h$, and all stage-frequencies are zero at times t_1 and t_n. For convenience, this will be assumed to be the case. If sampling times are not equally spaced then interpolation can be used to produce equally spaced data.

Let F_{ij} be the sample frequency for the number in stage j or a higher stage at time t_i. This is then the sample estimate of $F_j(t_i)$ in equations (4.1) to (4.3). The trapezoidal rule gives the area under the $F_j(t)$ curve to be estimated as

$$\hat{A}_j^* = h \sum_{i=1}^{n} F_{ij}, \qquad (4.10)$$

and the area under the $tF_j(t)$ curve to be estimated as

$$\hat{D}_j^* = h \sum_{i=1}^{n} t_i F_{ij}. \qquad (4.11)$$

It will be possible to specify times t_1 and t_n with zero stage-frequencies if sampling starts soon after individuals begin entering stage 1 and continues until virtually all individuals are dead, which is the first of the two conditions stated above for using the KNM method. For example, if Qasrawi's grasshopper data in Table 1.2 are considered then it seems fair enough to add 'samples' with zero frequencies on (say) 17 May and 28 September in order to use equations (4.10) and (4.11).

It is sometimes necessary to estimate the numbers entering stage 1 at different times. This can be done for equally spaced data using the equation

$$b_i = F_{i+1\,1} - \exp\{-\hat{\theta}(t_{i+1} - t_i)\} F_{i1}, \qquad (4.12)$$

where b_i is the estimated number entering stage 1 between the times of samples i and $i + 1$ that are still alive at the time of sample $i + 1$, F_{i1} is the frequency in all stages combined in the ith sample, and $\exp\{-\hat{\theta}(t_{i+1} - t_i)\}$ is the estimated probability of surviving the time t_i to t_{i+1}. The number entering the population between the ith and $(i + 1)$th sample times can be expected to be higher than b_i because some entries will die before a sample is taken. An adjusted estimate to take this into account is

$$b_i^* = b_i \theta(t_{i+1} - t_i)/\{1 - \exp(-\theta(t_{i+1} - t_i))\}. \qquad (4.13)$$

This follows from the fact that $\{1 - \exp(-\theta(t_{i+1} - t_i))\}/\{\theta(t_{i+1} - t_i))\}$ is the probability of an individual surviving until a sample is taken, if entry times to stage 1 are uniformly distributed between sample times.

The estimates produced from equations (4.12) and (4.13) can be negative. Some further adjustments may therefore be desired if estimates are to be used for generating expected stage-frequencies for simulation, or to compare with observed frequencies. For example, it may be sensible to replace any negative estimates by zero, and to adjust the positive ones so that they sum to the estimate of M_1 when they are added only for the period when individuals are present in stage 1.

Example 4.1   *Chorthippus parallelus* on East Budleigh Common

Qasrawi's grasshopper data in Table 1.2 can be used as a first example of the use of the KNM method. A simple check of the assumption of a constant survival rate is provided by plotting the logarithms of total sample sizes against time for the period after entries to stage 1 ceased. One implication of the survival assumption is that this plot should show an approximately linear decline, with the slope being the logarithm of the constant survival rate. Figure 4.1 indicates that this prediction does hold reasonably well for the grasshopper data and hence the assumption can be accepted, at least for the present. The slope is -0.029, corresponding to a daily survival rate of about 0.971.

To find the areas under the stage-frequency and time–stage-frequency curves, 'zero' samples were added to the data on 17 May and 28 September and sample frequencies at the equally spaced times $0, 4.5, 9.0, \ldots, 135$ were then obtained by interpolating between the unequally spaced times used by Qasrawi. The left-hand side of Table 4.3 shows the interpolated data obtained in this way. Note that there are the same number of equally spaced 'samples' as for the original data. The right-hand side of Table 4.3 shows the accumulated stage-frequencies to which equations (4.10) and (4.11) can be applied. From these accumulated frequencies it is found that the A^*, D^* and $B^* = D^*/A^*$ estimates are as follows:

j	$\hat{A}^*$	$\hat{D}^*$	$\hat{B}^*$
1	1055.7	51 729.3	49.0
2	776.9	46 380.9	59.7
3	543.4	37 603.3	69.2
4	406.5	31 219.2	76.8
5	285.8	24 493.1	85.7

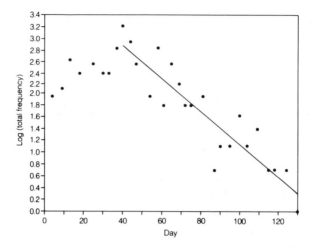

Figure 4.1   Test of the assumption of a constant survival rate for the grasshopper data of Table 1.2 by plotting the logarithms of total sample frequencies against time after entries to stage 1 cease. The line shown was fitted to the log frequencies after the peak by linear regression.

The corresponding estimates from equations (4.6) to (4.9) are shown below:

Stage	Daily survival rate	Stage-specific survival rate	Duration of stage (days)	Number entering stage
1	0.965	0.74	8.6	38
2	.	0.70	10.1	28
3	.	0.75	8.2	19
4	.	0.70	9.9	14
5	.	–	–	10

Note that the estimated daily survival rate is quite close to the value of 0.971 found from the slope of the linear regression line shown in Figure 4.1.

There are two obvious questions that arise at this point. First, how accurate can these estimates be expected to be? Second, how well does the KNM model fit the data?

A simple way to answer the first question involves simulating data similar to the observed data and seeing how much sampling variation occurs. This can be done fairly easily in the present case since the observed sample frequencies are (with two exceptions) counts that can reasonably be expected to have Poisson sampling errors. (The two exceptions are the first two

Table 4.3 Grasshopper data interpolated to give equally spaced sample times

	Stage frequencies					Cumulative stage frequencies				
Time	1	2	3	4	5	1	2	3	4	5
0	0.0					0.0				
4.5	7.1					7.1				
9.0	8.0					8.0	0.0			
13.5	13.6	0.1				13.7	0.1			
18.0	10.0	1.0				11.0	1.0	0.0		
22.5	8.1	3.6	0.6			12.3	4.2	0.6		
27.0	4.6	7.0	0.6			12.2	7.6	0.6	0.0	
31.5	1.0	9.0	0.5	0.5		11.0	10.0	1.0	0.5	
36.0	2.5	8.0	3.3	1.8		15.5	13.0	5.0	1.8	
40.5	6.1	11.4	6.0	0.7		24.3	18.1	6.7	0.7	0.0
45.0	0.3	5.0	6.0	5.3	0.3	17.0	16.7	11.7	5.7	0.3
49.5	0.6	1.0	4.9	3.3	1.0	10.9	10.2	9.2	4.3	1.0
54.0	0.0	1.0	3.0	2.0	1.0	7.0	7.0	6.0	3.0	1.0
58.5		3.3	3.5	3.8	4.5	15.2	15.2	11.8	8.3	4.5
63.0		0.5	1.0	4.0	4.0	9.5	9.5	9.0	8.0	4.0
67.5		1.0	1.0	3.1	5.4	10.5	10.5	9.5	8.5	5.4
72.0		0.0	0.0	0.0	6.0	6.0	6.0	6.0	6.0	6.0
76.5				0.2	6.0	6.2	6.2	6.2	6.2	0.2
81.0				1.0	6.0	7.0	7.0	7.0	7.0	6.0
85.5				1.0	2.2	3.2	3.2	3.2	3.2	2.2
90.0				0.0	3.0	3.0	3.0	3.0	3.0	3.0
94.5					3.0	3.0	3.0	3.0	3.0	3.0
99.0					4.6	4.6	4.6	4.6	4.6	4.6
103.5					3.2	3.2	3.2	3.2	3.2	3.2
108.0					3.8	3.8	3.8	3.8	3.8	3.8
112.5					2.8	2.8	2.8	2.8	2.8	2.8
117.0					2.0	2.0	2.0	2.0	2.0	2.0
121.5					2.0	2.0	2.0	2.0	2.0	2.0
126.0					1.7	1.7	1.7	1.7	1.7	1.7
130.5					0.9	0.9	0.9	0.9	0.9	0.9
135.0					0.0	0.0	0.0	0.0	0.0	0.0

frequencies, which are counts multiplied by 1.4. This minor complication will be ignored on the grounds that it can be expected to have very little effect on sampling variation.) A simulated set of data can be generated by replacing each observed stage-frequency with a random value from the Poisson distribution with mean equal to the observed value, which is what was defined above as a method four simulation in Section 4.3. For example, the observed stage-frequency of 14 on May 29 is replaced with a random value from the Poisson distribution with mean 14.

The results of simulating and analysing 100 sets of data in this manner are shown below:

Stage	Daily survival rate		Stage-specific survival rate		Duration of stage (days)		Number entering stage	
	Mean	SD	Mean	SD	Mean	SD	Mean	SD
1	0.964	0.001	0.73	0.03	8.9	1.3	35.1	3.5
2	.		0.68	0.03	10.5	1.4	25.5	2.8
3	.		0.74	0.04	8.2	1.5	17.4	1.9
4	.		0.70	0.05	9.7	1.6	12.9	1.4
5	.		–	–	–	–	9.1	1.0

The estimates of stage-specific survival rates and stage durations differ slightly from the original data estimates, indicating that there may be some minor biases in the estimation process for these parameters. More seriously, all of the simulation means for numbers entering stages are about 10% lower than the data estimates. These differences seem real. For example, the simulation mean for the number entering stage 1 (35.1) has a standard error of $3.5/\sqrt{100} = 0.35$. It is therefore about eight standard errors below the original data estimate of 38. If it was thought worthwhile, a bias correction could obviously be made. It seems that estimates of the numbers entering stages tend to be about 10% below true values. It is therefore appropriate to multiply estimates by 1.11 to make such a correction.

Type four simulations indicate only the variation in estimates due to sampling errors. The simulated sets of data will give a good indication of what can be expected from repeated sampling and estimation from the same population. However, the estimates obtained from the data using the KNM method may be quite different from the true values because the assumptions of the model are wrong. Therefore, to test the validity of the KNM model a different type of simulation is called for, using the KNM model itself to generate data. This is then a method three simulation according to the classification of Section 4.3.

Since the KNM model is so straightforward, using it to simulate data presents no particular problems. It is necessary to determine the numbers entering stage 1 at different times and project their development and survival forwards. The durations of stages are constant and survival rates can be applied deterministically to find expected stage-frequencies. In the simulations carried out with the example being considered, the numbers entering stage 1 in the time intervals 0–4.5, 4.5–9.0, 9.0–13.5, . . . were

Table 4.4 Estimated numbers entering the grasshopper population using equations (4.12) and (4.13)

Time interval	Estimated entries	Adjusted estimate	Used in simulations
0.0–4.5	7.1	7.7	6.9
4.5–9.0	1.9	2.1	1.9
9.0–13.5	6.9	7.4	6.7
13.5–18.0	−0.7	−0.7	0.0
18.0–22.5	2.9	3.2	2.8
22.5–27.0	1.7	1.9	1.7
27.0–31.5	0.6	0.7	0.6
31.5–36.0	6.1	6.6	6.0
36.0–40.5	11.0	11.9	10.8
40.5–45.0	−3.7	−4.0	0.0
45.0–49.5	−3.6	−3.9	0.0
49.5–54.0	−2.3	−2.4	
54.0–58.5	9.2	10.0	
58.5–63.0	−3.4	−3.7	
63.0–67.5	2.4	2.6	
67.5–72.0	−2.9	−3.2	
72.0–76.5	1.1	1.2	
76.5–81.0	1.7	1.8	
81.0–85.5	−2.7	−2.9	
85.5–90.0	0.2	0.2	
90.0–94.5	0.4	0.5	
94.5–99.0	2.0	2.2	
99.0–103.5	−0.7	−0.7	
103.5–108.0	1.0	1.1	
108.0–112.5	−0.4	−0.4	
112.5–117.0	−0.4	−0.4	
117.0–121.5	0.3	0.3	
121.5–126.0	−0.0	−0.0	
126.0–130.5	−0.5	−0.6	
130.5–135.0	−0.8	−0.8	
Total	34.7	37.5	37.5

estimated using equation (4.13) and then these were entered in five equal-sized batches over each of these intervals, after suitable adjustments to remove negative values and make the sum equal to the estimate of M_1. This 'spreading out' of entries is more realistic than entering all individuals at the same time, although the choice of five batches is admittedly arbitrary.

Estimated entry numbers are shown in Table 4.4. The column headed

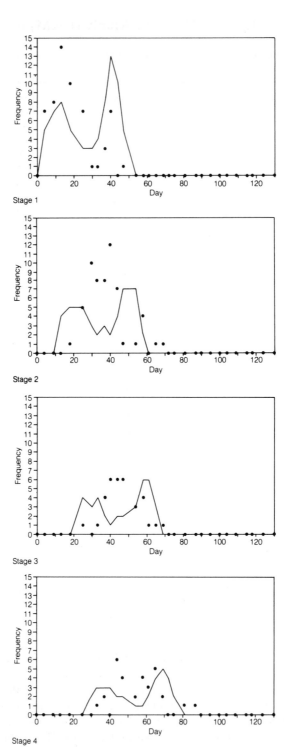

Stage 1

Stage 2

Stage 3

Stage 4

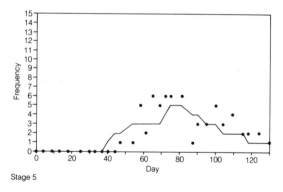

Stage 5

Figure 4.2   Comparison between the observed and expected frequencies for the KNM model fitted to the grasshopper data of Table 1.2. •, observed; —, expected.

'estimated entries' gives values from equation (4.12). In the next column are adjusted values from equation (4.13). The final column gives the values after a further adjustment involving replacing negative estimates by zero and making the positive estimates sum to 37.5 (the estimate of M_1). Estimates are only considered up to the time 49.5 because after then there were no individuals in stage 1, and hence presumably no entries.

Observed and expected frequencies are shown in Figure 4.2, this comparison being one indication of how well the KNM model fits the data. There are some obvious systematic discrepancies shown. For example, with stage 1 all the early observed stage-frequencies are above the expected values and all the late observed stage-frequencies are below the expected values.

Another way to examine the fit of the model to the data involves calculating a measure of the goodness of fit of the model to the data and seeing how this compares with the distribution of the goodness of fit found with simulated data. One possible measure is the sum of squares of differences between the observed sample stage-frequencies (f_{ij}) and the expected stage-frequencies calculated as just described ($f_j(t_i)$, say). That is, the goodness of fit is

$$G = \Sigma\{f_{ij} - f_j(t_i)\}^2, \qquad (4.14)$$

where the summation is over all sample stage-frequencies. This is better for the KNM model than a chi-squared statistic since by its nature the model can give a zero expected frequency when there is a non-zero observed frequency. This does not matter with a sum of squares, but will cause a division by zero with a chi-squared statistic.

If the observed goodness-of-fit statistic is significantly large in comparison with the distribution found for simulated data with the same parameters then it seems that the KNM model does not fit well. Hence, estimates of population parameters found using the method may be suspect. On the other hand, a poor fit may be due simply to extraneous sampling variation that can be accounted for by multiplying standard errors by a heterogeneity factor, as discussed in Sections 2.11 and 3.5. This heterogeneity factor, which is the ratio of the observed sampling variances to the sampling variances expected with Poisson sampling errors, can be estimated by the ratio of the observed value of G to the mean value from data simulated with the same entry numbers, stage durations and survival rate as estimated from the observed data.

The value of G for the real data is 837.8. When 100 sets of data were generated with the estimated entry numbers up to time 40.5 shown in Table 4.4, the goodness-of-fit statistics obtained ranged from 163 to 648. Thus the goodness-of-fit value for the real data is rather larger than is expected from the KNM model, being significant at the 1% level or more.

The simulations give a mean G value of 336.6, which corresponds approximately to the expected value if sampling errors in stage-frequencies are Poisson distributed. A heterogeneity factor can therefore be estimated as $H = 837.8/336.6 = 2.49$. The standard errors given above that were determined by a type four simulation, assuming Poisson-distributed sampling errors, can be 'corrected' for this factor by multiplying by $\sqrt{H} = 1.58$ on the assumption that it is excess sampling variation causing the poor fit of the model. Of course, the reason for the large observed G value may simply be that the model is not appropriate. The patterns in the differences between observed and expected frequencies shown in Figure 4.2 suggest that this is likely to be the case.

4.7 THE KNM METHOD WITH ITERATIVE CALCULATIONS

If the first sample from a population shows many individuals in stage 2 or a higher stage, or the last observation includes many individuals still alive, then the simple expedient of adding zero samples at the start and end of the sampling period will not be satisfactory. In such cases an iterative method for adjusting the data may work (Manly, 1985b). The procedure is as follows:

1. Estimates are calculated from the available data, using equations (4.6) to (4.11), ignoring the requirement that stage-frequencies should be zero in the first and last samples.
2. The $\hat{A}_j^*$ and $\hat{D}_j^*$ values of equations (4.10) and (4.11) are corrected to allow for non-zero first and/or last samples using the equations

$$\hat{A}_j' = \hat{A}_j^* + \left\{ F_{nj} - F_{1j} + \sum_{k=1}^{j-1} (f_{nk} - f_{1k}) \exp(-\theta l_{kj}) \right\} / \theta,$$

and

$$\hat{D}'_j = \hat{D}^*_j + \{F_{nj}(t_n \theta + 1) - F_{1j}(t_1 \theta + 1)\}/\theta^2$$

$$+ \sum_{k=1}^{j-1} \exp(-\theta l_{kj})[f_{nk}\{(t_n + l_{kj})\theta + 1\} - f_{1k}\{t_1 + l_{kj})\theta + 1\}]/\theta^2.$$

Here $l_{kj} = a_k/2 + a_{k+1} + \ldots + a_{j-1}$ is the average time for an individual in stage k to enter stage j, when $j < k$ (assuming that they are mid-way through the stage to begin with). Estimates are used as necessary in place of unknown population values in these equations.

3. Estimates of population parameters are recalculated using the corrected values $\hat{A}'$ and $\hat{D}'$.
4. If the recalculated estimates are appreciably different from the previous-ly calculated values then a return is made to step (2); otherwise iteration stops.

Experience has indicated that this iterative approach to estimation usually works with data that are incomplete due to sampling terminating too soon. However, convergence problems are apt to occur when sampling begins too late, so that some individuals have passed out of stage 1 by the time of the first sample.

Example 4.2   Analysis of the Incomplete *Chorthippus parallelus* Data

As an example of the use of the iterative method, suppose that the last sample from Qasrawi's grasshopper population was taken on 5 August, at which time all individuals had not reached the adult stage (Table 1.2). In this case the estimates by iteration are as follows:

Stage	Daily survival rate	Stage-specific survival rate	Duration of stage (days)	Number entering stage
1	0.983	0.78	14.0	23
2	.	0.77	14.7	18
3	.	0.82	11.0	14
4	.	0.83	10.3	11
5	.	–	–	9

A comparison with the estimates from the full data that are given above shows some substantial differences. For the partial data the daily and stage-specific survival rates are higher, the mean stage durations are longer, and the numbers entering stages are lower.

As for full data, simulation can be used to determine how accurate the estimates from the partial data are. To this end, 100 sets of data were generated by replacing the observed stage-frequencies with random values from Poisson distributions, with these Poisson distributions having means equal to the observed stage-frequencies (type four simulation, Section 4.5). The means and standard deviations are as shown below:

	Daily survival rate		Stage-specific survival rate		Duration of stage (days)		Number entering stage	
Stage	Mean	SD	Mean	SD	Mean	SD	Mean	SD
1	0.979	0.012	0.78	0.07	14.0	4.3	23.5	8.5
2	.		0.77	0.09	14.4	3.3	17.8	5.1
3	.		0.81	0.09	10.4	1.9	13.3	2.5
4	.		0.81	0.11	10.3	2.0	10.6	1.1
5	.		–	–	–	–	8.5	1.1

The partial data estimates are considerably less accurate than the full data ones. Interestingly, though, there is less evidence of bias apparent here than was the case for the full data estimates.

Although this is only one example, it does indicate that if the KNM method of estimation is to be used then every effort should be made to sample the entire development period.

4.8   THE KEMPTON METHOD OF ESTIMATION

The first real stochastic model for stage-frequency data was provided by Read and Ashford (1968), and discussed further by Ashford *et al.* (1970). They assumed an Erlangian distribution for the times spent in different stages and determined maximum likelihood estimates of population parameters for the grasshopper data of Table 1.2, and another set of similar data. Unfortunately, the particular distributional assumptions made by Read and his colleagues resulted in equations that are not particularly suitable for computational purposes. Since distributional assumptions are to a large extent arbitrary, this is something of a drawback.

Kempton (1979) was able to overcome these problems by modifying the assumptions slightly. He began his paper with a review of the factors that have to be considered in developing a proper stochastic model for a population passing through several developmental stages. He then went on to discuss the fitting of models to data using the principle of maximum likelihood, and gave two examples. Here his general comments will be briefly summarized and then one of his models, based on the gamma distribution, will be considered in more detail.

Kempton noted that models must include three components, the first of which concerns the survival rates in different stages. He suggests that a constant survival rate per unit time may be too simplistic, although this may be an acceptable assumption within one life stage. Assuming a gamma, Weibull or lognormal distribution for lengths of life may be more satisfactory. The second component of a model concerns times of entry to stage 1. For this, Kempton suggested that it is realistic to conceive of development starting at a time 0 with all individuals in a stage 0. The time of entry to stage 1 for an individual then corresponds to the duration of stage 0 for that individual. A normal or a gamma distribution may be realistic for the distribution of this duration. Finally, the third component of a model concerns the durations of stages $1, 2, \ldots, q$. In this case, Kempton pointed out the advantages of assuming that durations are normally distributed, gamma distributed, inverse normally distributed, or constant for all individuals.

As far as the gamma distribution is concerned, the advantages of its use come about because of an 'additive' property. The gamma distribution with scale parameter b and shape parameter k has the probability density function

$$g(x) = b^k x^{k-1} \exp(-bx)/\Gamma(k), \qquad b, k > 0,$$

which can conveniently be referred to as the $G(k, b)$ distribution. The mean is $\mu = k/b$, and the variance is k/b^2.

Suppose that the time of entry to stage 1 follows a $G(k_0, b)$ distribution, and the duration of stage j follows a $G(k_j, b)$ distribution, for $j = 1, 2, \ldots, q - 1$, with all these gamma distributions being independent. Then it can be shown that the time of entry to stage j will follow a $G(c_j, b)$ distribution, where

$$c_j = k_0 + k_1 + \ldots + k_{j-1}.$$

The mean time of entry to stage j will therefore be $\mu_j = c_j/b$. The mean duration of stage j is $a_j = k_j/b$, so that the equation

$$\mu_j = \mu_1 + a_1 + a_2 + \ldots + a_{j-1}$$

also applies.

A gamma distribution for the time of entry to stage 1 may well be a realistic approximation for situations where there is a single peak in the curve of the numbers entering the population. Also, a gamma distribution for the duration of a life stage is a reasonable assumption in the absence of any knowledge about the true form of the distribution. However, it seems difficult to justify the constant scale parameter b on any grounds other than mathematical convenience. Indeed, Read and Ashford (1968) and Ashford *et al.* (1970) used a gamma model for which k values were constant between stages and b values varied. (The Erlangian distribution that they used is a special case of the gamma distribution with an integer value of k.)

Whatever particular assumptions are made about the form of distributions, the probability that an individual is in stage j at time t can be written as

$$p_j(t) = w(t) \int_0^t \{g_j(y) - g_{j+1}(y)\} dy, \qquad j < q, \qquad (4.15)$$

and

$$p_q(t) = w(t) \int_0^t g_q(y) dy, \qquad (4.16)$$

where $w(t)$ is the probability of surviving to time t, and $g_j(t)$ is the probability density function of the time of entry to stage j. Here time t is measured relative to the process starting time of zero; the survival rate function $w(t)$ applies for individuals in stage 0 as well as individuals in other stages. The integral part of the equation gives the probability that a survivor to time t is in stage j: the integral from 0 to t of $g_j(y)$ gives the probability of having entered stage j before time t while the integral of $g_{j+1}(y)$ gives the probability of having entered stage $j + 1$ before this time. The difference between the integrals of $g_j(y)$ and $g_{j+1}(y)$ is therefore the probability of having entered but not left stage j.

There are some implicit assumptions involved in the use of equations (4.15) and (4.16). It is assumed that the probability of an individual surviving depends only on its age, and not on its developmental rate. It is also assumed that the times that an individual spends in different stages are independently distributed. Thus, if an individual passes quickly through stage 1 then this is assumed not to provide any information about the time needed to pass through later stages. In practice, it seems likely that the times that an individual spends in different stages will be related. Fast and slow developers probably exist. However, as Kempton has noted, the data available for fitting models will seldom allow this type of complication to be taken into account.

The parametric forms of the functions $w(t)$ and $g_j(t)$ will determine how easy it is to compute values for $p_j(t)$. For his numerical examples, Kempton assumed that the survival per unit time was constant and took $w(t) = \exp(-\theta t)$, which is the simplest possible realistic function. Kempton also assumed gamma distributions with a common scale parameter b for the time of entry to stage 1 and the durations of stages, as discussed above. Thus the function $g_j(y)$ of equations (4.15) and (4.16) becomes the probability density function of a gamma distribution with scale parameter b and shape parameter

$$c_j = b(\mu_1 + a_1 + a_2 + \ldots + a_{j-1}). \tag{4.17}$$

The integral part of equation (4.15) then consists of the difference between the cumulative probability functions for the two gamma distributions $G(c_j, b)$ and $G(c_{j+1}, b)$, where these functions can be evaluated using the algorithm proposed by Lau (1980). Similarly, the integral part of equation (4.16) is the cumulative distribution function of the $G(c_q, b)$ distribution, which can be evaluated using the same algorithm.

The gamma model was fitted to two sets of data by Kempton, one of these sets being that provided in Table 1.2. For the fitting process he assumed that the sample count of the number of individuals in stage j in the sample at time t_i was a random value from the Poisson distribution with mean $M_0 p_j(t_i)$, where M_0 is the total number of individuals at time 0. Kempton provides the equation for the likelihood function under these conditions. He maximized it numerically using a two-stage procedure with a general purpose maximum likelihood program. An alternative fitting process is possible using the computer program MAXLIK that has been discussed in the previous chapter. If the data stage-frequencies f_{ij} are Poisson distributed with means $M_0 p_j(t_i)$, as suggested by Kempton, then the distribution of the stage-frequencies conditional on the observed total frequency in all samples is multinomial. The probability of an observation occurring in stage j in the sample at time t_i is then

$$P_{ij} = p_j(t_i) / \sum_{u=1}^{q} \sum_{v=1}^{n} p_u(t_v),$$

for $j = 1, 2, \ldots, q$, and $i = 1, 2, \ldots, n$.

With this multinomial model the parameters to be estimated are b, the scale parameter for the gamma distributions; μ_1, the mean time of entry to stage 1; θ, the survival parameter; and $a_1, a_2, \ldots, a_{q-1}$, the durations of stages 1 to $q - 1$. This is assuming that the time the process starts is known so that the sample times $t_1, t_2, \ldots, t_n$ can be taken relative to this. If the starting time is not known it can be included as a parameter t_0 and sample times taken as $t_i - t_0$. However, this is likely to result in an overparameterized model, so it is probably best to fix the starting time at some realistic point before the first sample. Some numerical results provided by Munholland (1988) suggest that the choice of a starting time will not be too crucial as far as estimates of stage durations and survival rates are concerned.

Inspection of the data should suggest sensible approximations for θ (the logarithm of the unit-time survival rate), μ_1, and a_1 to a_{q-1} to start the estimation process. A suitable value for b usually seems to be 1.

Once the mean stage durations and other parameters have been estimated with the multinomial model, they can be used to determine estimates of other population parameters. If N_0 is the total number of individuals counted in all stages in all samples then this is related to M_0, the number of individuals in the population at time zero, by

$$N_0 \simeq M_0 \sum_{i=1}^{n} \sum_{j=1}^{q} p_j(t_i).$$

Hence an obvious estimator of M_0 is

$$\hat{M}_0 = N_0 / \sum_{i=1}^{n} \sum_{j=1}^{q} p_j(t_i). \tag{4.18}$$

The number of individuals entering stage j is the number of the M_0 that survive that long, where the required survival time is a gamma $G(c_j, b)$ random variable. This is therefore the average value of the survival function, which is given by the equation

$$M_j = M_0 \int_0^\infty \exp(-\theta t) g_j(t) \, dt$$

$$= M_0 \int_0^\infty \exp(-\theta t) b^{c_j} t^{c_j - 1} \exp(-bt) \, dt / \Gamma(c_j)$$

$$= M_0 (1 + \theta/b)^{-c_j}, \tag{4.19}$$

where c_j is given by equation (4.17). Substituting estimates of θ, b, μ_1 and the mean stage durations into the right-hand side of this last equation produces estimates $\hat{M}_1$, $\hat{M}_2$, ..., $\hat{M}_q$ of the numbers entering stages. Stage-specific survival rates are then estimated using the obvious equation

$$\hat{w}_j = \hat{M}_{j+1} / \hat{M}_j. \tag{4.20}$$

The estimated unit time survival rate is related to the estimate $\hat{\theta}$ of θ as $\hat{\phi} = \exp(-\hat{\theta})$. There is also a simple approximate relationship between variances, with

$$\text{var}(\hat{\phi}) \simeq \phi^2 \text{var}(\hat{\theta}), \tag{4.21}$$

this being obtained by a Taylor series expansion.

Kempton actually had two gamma distribution models. One of them was a model with stage durations being constant for all individuals. This leads to some minor changes in the above equations but does not reduce the number of parameters. His second model differs from the one described here in one respect only: he included a parameter a_q for the duration of stage q and made equation (4.15) apply for $j = q$ as well as $j < q$. This amounts to assuming that losses from the final stage are due to more than just deaths, which is sensible if there is adult emigration, for example.

In principle, there is no difficulty in estimating an a_q parameter using the multinomial maximum likelihood estimation model that has been described above. However, in practice this seems to increase considerably convergence problems with the iterative numerical process. Kempton mentions convergence problems, without giving any details. It seems likely that for many sets of data the incorporation of the extra parameter will result in a model that is overparameterized, symptoms of which are convergence difficulties because of highly correlated parameter estimators. In short, it seems best to avoid including a duration for stage q in models unless there are good reasons for the inclusion.

If developmental rates are temperature dependent then physiological time (e.g., degree days above a threshold level) can be used in place of calendar time in the above equations. Alternatively, physiological time can be used with the gamma distributions for stage durations and the time of entry to stage 1, but calendar time can be used in the survival function. To do this requires only minor changes to most of the equations given above. Equations (4.15) and (4.16) become

$$p_j(t) = w(t) \int_0^{t'} \{g_j(y) - g_{j+1}(y)\} dy, \quad j < q, \tag{4.22}$$

and

$$p_q(t) = w(t) \int_0^{t'} g_q(y) dy, \tag{4.23}$$

where t' is the physiological time corresponding to calendar time t. The only real complication is with equation (4.19). In this case

$$M_j = M_0 \int_0^{\infty} \exp(-\theta t) g_j(t') dt, \tag{4.24}$$

can no longer be integrated analytically. The only approach to evaluating an estimate of M_j then seems to be to determine the integral numerically.

Example 4.3   Analysis of the Incomplete *Chorthippus parallelus* Data by Kempton's Method

As an example of the use of Kempton's method, consider again Qasrawi's grasshopper data shown in Table 1.2 up to 5 August, assuming that individuals started entering stage 1 on day 0 (May 16). Maximum likelihood estimates of the parameters used with the MAXLIK algorithm are shown in the following table, together with estimated standard errors. The major discrepancy between these estimates and those from the KNM method is for the duration of stage 1, which was estimated at 14.0 days for the KNM method.

Parameter	Estimate	Standard error
Scale, b	0.343	0.051
Mean time of entry to stage 1, μ_1	6.677	1.567
Survival parameter, θ	0.009	0.004
Duration of stage 1, a_1	21.349	2.006
Duration of stage 2, a_2	15.892	1.711
Duration of stage 3, a_3	9.711	1.580
Duration of stage 4, a_4	12.256	1.901

The daily survival rate is estimated as $\exp(-\hat{\theta}) = 0.991$, with standard error 0.004 (equation (4.21)). This compares with the KNM estimate of 0.983. The estimated numbers entering stages 1–5 are found using equation (4.19) to be $\hat{M}_1 = 17.1$, $\hat{M}_2 = 14.2$, $\hat{M}_3 = 12.3$, $\hat{M}_4 = 11.3$ and $\hat{M}_5 = 10.2$. These are in reasonable agreement with the KNM estimates.

The deviance of the model (minus twice the maximized log-likelihood) is 82.33 with 50 degrees of freedom. Compared with the percentage points of the chi-squared distribution this is significantly large at the 1% level. The conclusion must be, therefore, that the fit of the Kempton model is not particularly good with these data. However, a graphical comparison of the observed and the corresponding expected stage-frequencies calculated from the model (Figure 4.3) indicates no particular patterns in the deviations from the model. This means that it is realistic to assume a constant heterogeneity factor H and adjust the standard errors accordingly, as discussed in Section 3.6. The estimate of H is the deviance divided by the degrees of freedom, which is $82.33/50 = 1.65$. The standard errors shown above should therefore be adjusted by multiplying by $\sqrt{1.65} = 1.28$.

Example 4.4   Analysis of a Generation of *Orchesella cincta*

As a second example of Kempton's method, consider the data in Table 4.1 for a generation of the litter-inhabiting *Orchesella cincta* in a forest near Dronten, The Netherlands. Recall that here the 'stages' are seven size classes with the lower limits 0.73, 0.90, 1.35, 1.80, 2.25, 2.70 and 3.15 mm.

Figure 4.4 indicates that the fall-off in total numbers after all entries to stage 1 ceased is consistent with the assumption of a constant survival rate so that Kempton's model is potentially applicable with a single survival parameter. The results that follow are based on using calendar time for development, with a starting time of week 0.

The parameter estimates obtained from the MAXLIK algorithm are shown below, together with estimated standard errors. The estimated survival rate per unit time is $\hat{\phi} = \exp(-0.206) = 0.813$, with standard error 0.005. The estimated numbers entering stages 1–7 are 1524.6, 1015.0, 686.1, 454.6, 270.8, 137.2 and 34.4, respectively.

Parameter	Estimate	Standard error
Scale, b	0.921	0.029
Mean time of entry to stage 1, μ_1	3.987	0.094
Survival parameter, θ	0.206	0.005
Mean duration of stage 1, a_1	2.183	0.066
Mean duration of stage 2, a_2	2.101	0.067
Mean duration of stage 3, a_3	2.210	0.084
Mean duration of stage 4, a_4	2.779	0.117
Mean duration of stage 5, a_5	3.649	0.180
Mean duration of stage 6, a_6	7.390	0.450

These estimates appear reasonable, with fairly small standard errors. However, the fit of the model to the data is very poor since the deviance is 564.6 with 48 degrees of freedom, giving a heterogeneity factor of $H = 11.76$. With such a poor fit it is questionable whether simply adjusting standard errors with a heterogeneity factor to allow for excess sampling errors is sensible, although the comparison between observed and expected frequencies shows no particular patterns in deviations other than the last three observed frequencies being low in stage 7 (Figure 4.5). Still, it may be that the model is wrong. For example, daily survival rates may have varied from stage to stage or the distributions of stage durations may not be well approximated by gamma distributions with a constant scale parameter. If the model is accepted then the standard errors given above should all be multiplied by $\sqrt{H} = 3.43$.

Another possibility is that the development through stages may have depended on physiological time rather than calendar time. Van Straalen (1985) discussed the determination of physiological time, and provided the values for this shown in Table 4.1.

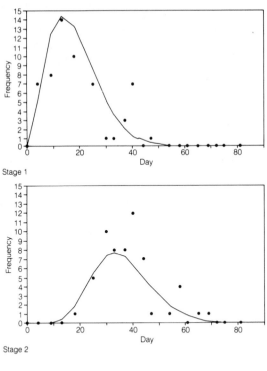

Stage 1

Stage 2

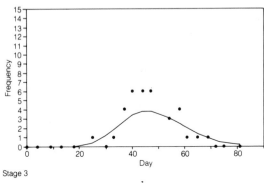

Stage 3

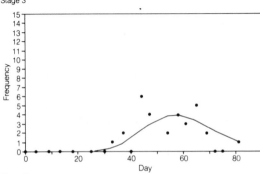

Stage 4

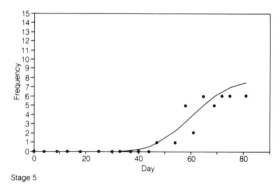

Stage 5

Figure 4.3 Comparison between the observed and expected stage-frequencies for Kempton's model fitted to the partial grasshopper data from Table 1.2. ●, observed; —, expected.

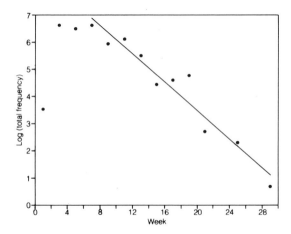

Figure 4.4 Test of the assumption of a constant survival rate for the *Orchesella cincta* data of Table 4.1 by plotting logarithms of total sample frequencies against time after entries to stage 1 cease.

4.9 VARIATIONS OF THE KEMPTON TYPE OF MODEL

As mentioned in the last section, Kempton (1979) noted that there is mathematical convenience with assuming normally distributed, gamma distributed, inverse normally distributed, or constant stage durations. In fact, the theory for the gamma distribution is extended fairly easily to the normal and inverse normal distributions. The equations (4.15) and (4.16) apply immediately with the probability density functions changed appropriately. Equations (4.18), (4.20) and (4.21) still hold, and equation (4.19) only needs obvious changes.

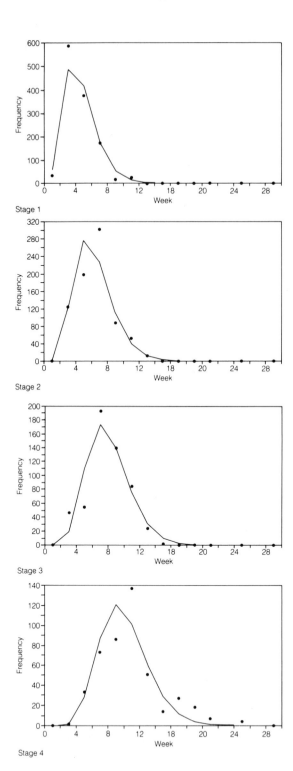

Stage 1

Stage 2

Stage 3

Stage 4

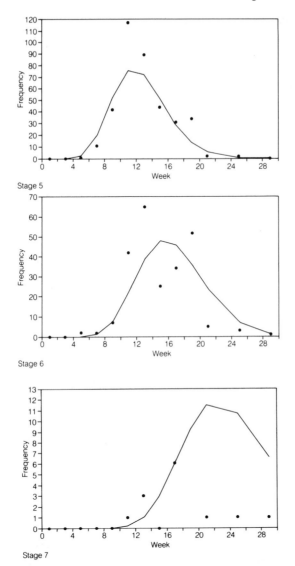

Figure 4.5 Comparison of observed and expected stage-frequencies for the Kempton model fitted to the *Orchesella cincta* data of Table 4.1. Where ●, observed; —, expected.

Before considering the details of changes in the assumptions concerning stage durations, another possible variation in the model can be noted. This concerns the survival function $w(t)$ in equations (4.15) and (4.16). Although taking this as $\exp(-\theta t)$ will often be adequate, a more complicated function may be considered. For example, the survival function

$$w(t) = \exp(-\theta_1 t - \theta_2 t^2) \qquad (4.25)$$

could be used to account for a trend in the daily survival rate. The equation (4.19) for the numbers entering stages with the gamma distribution model then no longer applies. Numerical integration would have to be used to evaluate

$$M_j = M_0 \int_0^\infty w(t) g_j(t) \, dt. \qquad (4.26)$$

However, equations (4.15) and (4.16) still apply. Another possibility is to have a survival function that changes at a known point in time because of some specific change in the circumstances of the population, such as the application of an insecticide.

Suppose that a normal distribution is assumed for the time of entry to stage 1, and that the durations of stages are also normally distributed. In that case the distributions of the times of entry to the other stages will also be normally distributed. The probability density function for the entry time to stage j needed for equations (4.15) and (4.16) will then have the form

$$g_j(x) = \exp[-\{(x - \mu_j)/\sigma_j\}^2]/\{\sqrt{(2\pi)}\,\sigma_j\}. \qquad (4.27)$$

Here the mean μ_j and variance σ_j^2 will be related to the mean and variance of stage durations by the equations

$$\mu_j = \mu_1 + a_1 + a_2 + \ldots + a_{j-1}, \qquad (4.28)$$

and

$$\sigma_j^2 = \sigma_1^2 + s_1^2 + s_2^2 + \ldots + s_{j-1}^2, \qquad (4.29)$$

where a_j is the mean and s_j is the standard deviation of the duration of stage j.

To reduce the number of parameters to be estimated it might be appropriate to set the standard deviations of stage durations to a constant, and take

$$\sigma_j^2 = \sigma_1^2 + (j-1)s^2. \qquad (4.30)$$

In some cases, the model will then fit the data best with a negative value for s^2. This indicates that variation is being reduced as individuals pass through stages, which does seem to be a real phenomenon with some populations, although the reason for its occurrence is unclear.

One interesting aspect of the normal distribution model is the existence of an explicit equation for the number entering stage j even when the survival function is the two-parameter one given by equation (4.25). Equation (4.26) becomes for this model

$$M_j = M_0 \exp[\{\theta_1(\tfrac{1}{2}\theta_1 \sigma_j^2 - \mu_j) - \theta_2 \mu_j^2)\}/\{1 + 2\theta_2 \mu_j^2\}]. \qquad (4.31)$$

The normal distribution model for stage durations has been discussed recently by Braner (1988) and Braner and Hairston (1989). They note that this model can be used even for cases where survival rates vary from stage to stage. However, this is not a generalization that will be considered here.

The use of the inverse normal (sometimes called the inverse Gaussian) distribution for modelling stage durations has been discussed at length by Munholland (1988); see also Kemp et al. (1989) and Munholland et al. (1989). This model can be justified in terms of a diffusion process and a development variable. The description 'inverse normal' comes about because the reciprocals of stage durations are normally distributed.

If it is assumed that the distribution of the time of entry to stage 1 and the durations of stages 1 to $q - 1$ are inverse normal with the same scale factor q, then the probability density function for the time of entry to stage j takes the form

$$g_j(x) = \sqrt{\{Q_j/(2\pi x^3)\}} \exp\{- Q_j(x - \mu_j)^2/(2\mu_j^2 x)\}, \qquad (4.32)$$

Q_j and μ_j being stage-dependent parameters. The diffusion model also implies that the parameter $B = \sqrt{Q_j}/\mu_j$ is constant, so that $Q_j = (B\mu_j)^2$.

Substituting equation (4.32) into equations (4.15) and (4.16) allows the proportions in different stages to be calculated at different times. The calculations are aided by noting that the cumulative distribution function (the probability of a value less than x) is

$$G_j(x) = \Phi(u_j) + \exp(2Q_j/\mu_j)\Phi(-v_j), \qquad (4.33)$$

where $\Phi(u)$ is the cumulative distribution function for the standard normal distribution, $u_j = (Q_j/x)(x/\mu_j - 1)$ and $v_j = (Q_j/x)(x/\mu_j + 1)$.

With this model the equation (4.28) still gives the relationship between the mean durations of stages (a_j) and the mean times of entry (μ_j). Since the scale factor is fixed, the variance of the time of entry to stage j is $\sigma_j^2 = Q_j^3/\mu_j$, while the variance of the duration of stage j is $s_j^2 = Q_j^3/a_j$.

If the survival function is $w(t) = \exp(-\theta t)$ then there is the explicit expression

$$M_j = M_0 \exp[\mu_j B\{-(B^2 + 2\theta)\}] \qquad (4.34)$$

for the number entering stage j. If the two-parameter function (4.25) is used then numerical integration will have to be used to evaluate the right-hand side of equation (4.26).

Example 4.5   Further Analysis of the *Orchesella cincta* Data

If the *Orchesella cincta* data used in Example 4.4 are analysed, assuming either normal or inverse normal distributions for stage durations, then the results are rather similar to what has been found for Kempton's gamma model. Table 4.5 shows how parameter estimates compare for the three models. Estimated standard errors are also given in the table. Note, however, that none of the models fits the data well. Hence, at the very least, these standard errors should be adjusted by heterogeneity factors before they can be taken seriously.

Table 4.5 Comparison of parameter estimates and deviances with three models fitted to the *Orchesella cincta* data

Parameter	Gamma model Est.	SE	Inverse normal model Est.	SE	Normal model Est.	SE
Mean time of entry to stage 1	3.987	0.094	3.896	0.097	2.864	0.064
Variance of time of entry	–	–	–	–	1.096	0.079
Survival parameter	0.206	0.005	0.199	0.005	0.178	0.004
Mean duration of stage 1	2.183	0.066	2.350	0.067	2.664	0.061
Mean duration of stage 2	2.101	0.067	2.140	0.070	2.372	0.066
Mean duration of stage 3	2.210	0.084	1.739	0.088	2.507	0.086
Mean duration of stage 4	2.779	0.117	2.973	0.126	3.022	0.118
Mean duration of stage 5	3.649	0.180	3.827	0.192	3.853	0.177
Mean duration of stage 6	7.390	0.450	7.748	0.471	7.306	0.408
Variance of stage durations	–	–	–	–	3.370	0.133
Number entering stage 1	1525	–	1482	–	1324	–
Number entering stage 2	1015	–	980	–	868	–
Number entering stage 3	686	–	672	–	600	–
Number entering stage 4	455	–	495	–	405	–
Number entering stage 5	271	–	293	–	249	–
Number entering stage 6	137	–	150	–	132	–
Number entering stage 7	34	–	38	–	38	–
Gamma scale parameter	0.921	0.029	–	–	–	–
Inverse normal scale parameter	–	–	0.817	0.017	–	–
Deviance of fitted model	564.6		639.1		714.0	
Degrees of freedom	48		48		47	

Deviances are given at the foot of Table 4.5. Although none of the models fits the data well, it does appear that the gamma model is somewhat better than the other two models in terms of this measure of goodness of fit.

4.10   THE BELLOWS AND BIRLEY MODEL

Bellows and Birley (1981) extended a discrete time model that was proposed by Birley (1977, 1979) and developed equations for estimation with multi-cohort data that allow for a different survival rate for each stage but need a knowledge of the distribution of times of entry to stage 1 in addition to sample stage-frequencies. If, as is usually the case, the entry distribution is not known, then it can be estimated by assuming that it has a suitable parametric form.

Although Bellows and Birley provided equations only for the development of the individuals in a single cohort entering stage j at the same time, it is fairly straightforward to extend these equations to cover the development of several cohorts entering a stage at different times. The first important equation is

$$f_j(t) = M_1 \sum_{i=0}^{t} P_j(i) \phi_j^{t-i} \{1 - H_j(t-i)\}, \qquad (4.35)$$

which says that the frequency in stage j at time t consists of contributions from those entering at times $0, 1, 2, \ldots, t$. The number entering at time i is the total number of individuals entering stage 1, M_1, multiplied by the proportion of these entering stage j at this time, $P_j(i)$. These individuals will still be in the stage at time t if they survive, with probability ϕ_j^{t-i}, and do not pass into stage $j + 1$, with probability $1 - H_j(t - i)$. Thus, the contribution to $f_j(t)$ from entries at time i is given by the ith term in the summation on the right-hand side of the equation.

In the above equation, $H_j(t)$ denotes the probability that an individual spends t or less units of time in stage j. Hence, if $h_j(t)$ denotes the probability of a stage duration of t time units, then $H_j(t) = h_j(0) + h_j(1) + \ldots + h_j(t)$.

A second important equation is

$$P_{j+1}(t) = \sum_{i=0}^{t} P_j(i) \phi_j^{t-i} h_j(t-i), \qquad (4.36)$$

which shows the proportion of the population entering stage $j + 1$ at time t as a function of the proportions entering stage j at different times. It says that the entries to stage $j + 1$ at this time is made up of individuals entering stage j at times $0, 1, 2, \ldots, t$ that survive to leave stage j at time t. Thus, the ith term in the summation is the proportion entering stage j at time i, $P_j(i)$, multiplied by the probability of surviving the time interval i to t, ϕ_j^{t-i}, multiplied by the probability of a stage duration of $t - i$, which is $h_j(t - i)$.

To complete the model it is necessary to make some assumptions about the distributions of durations of stages, and the time of entry to stage 1. Bellows and Birley suggest that it is realistic to assume that the developmental rate in a stage is normally distributed between individuals so that the distribution of the duration of a stage is inverse normal. This reflects the rather skewed distributions that have been observed experimentally. Obviously, there are other possibilities. Here the Weibull distribution is proposed on the grounds that the cumulative distribution function takes the particularly simple form

$$H_j(t) = 1 - \exp\{-(t/Q_j)^{\alpha_j}\}. \tag{4.37}$$

The Weibull is also convenient for modelling the distribution of the time of entry to stage 1. For this

$$P_1(t) = \exp[-\{(t-1)/Q_0\}^{\alpha_0}] - \exp[-(t/Q_0)^{\alpha_0}], \tag{4.38}$$

for $i = 1, 2, \ldots, t$ with $P_1(0) = 0$. In these equations Q_0 and Q_j are scale parameters to be estimated.

Bellows and Birley suggested using non-linear regression to estimate the parameters of their model. However, if the data stage-frequencies are regarded as Poisson variates then the MAXLIK algorithm can be used instead. Equations (4.35) to (4.38) then have to be used to determine the proportion of all stage-frequencies expected in stage j at time t, which is

$$p_j(t) = f_j(t)/ \sum_{i=1}^{n} \sum_{j=1}^{q} f_j(t),$$

assuming samples are available for times $1, 2, \ldots, n$.

To start the estimation process using MAXLIK or some other iterative technique, the α values can be set equal to 1. In that case Q_0 is the mean time of entry to stage 1 and the other Q values are mean durations of stages. Inspection of the data should produce sensible values for these parameters and the unit time survival rate.

This method of estimation has some possible advantages over Kempton's method and its variations. Two parameters (α and Q) can be used to define the distribution of the duration of a stage. Also, a different survival rate in each stage is possible. Of course, using this flexibility may produce an overparameterized model. In practice, it may therefore be best to make either the α or the Q values (or both) the same in each stage.

If a population 'catastrophe' (e.g., insecticide spraying) occurs at a known point in time then it is possible to apply either a special instantaneous mortality to all stages, or to allow the survival rates to change permanently from that time on (Bellows et al., 1982).

Once the basic parameters of Bellows and Birley's model have been estimated using the MAXLIK algorithm, or some other method, other secondary parameters can be determined quite easily. Since $P_j(i)$ is the proportion of the population entering stage j at time i,

$$w_j = \sum_{i=1}^{n} P_{j+1}(i) / \sum_{i=1}^{n} P_j(i) \qquad (4.39)$$

is the stage-specific survival rate for stage j, where n is a large enough time so that all individuals have left stage j. Substituting estimated values into the right-hand side of this equation therefore produces estimates of these survival rates. Obviously, sampling has to continue long enough for $P_{j+1}(n)$ to be estimable. The number entering stage j is given by

$$M_j = M_1 w_1 w_2 \ldots w_{j-1}. \qquad (4.40)$$

This can be estimated by substituting estimated stage-specific survival rates in the right-hand side of the equation and estimating M_1 by

$$\hat{M}_1 = \sum_{i=1}^{n} \sum_{j=1}^{q} f_{ij} / \sum_{i=1}^{n} \sum_{j=1}^{q} f_j(i), \qquad (4.41)$$

the ratio of the total observed frequency to the total of the probabilities of being in different stages at different times.

Finally, it can be noted that although this model has been defined for stage-frequencies counted at equal intervals of time, this restriction does not necessarily have to apply. It is quite possible to generate population probabilities on this basis and then interpolate between them to get expected frequencies corresponding to non-integer sample times.

Example 4.6   Bellows and Birley's Method Applied to the Full *Chorthippus parallelus* Data

The Bellows and Birley model can be applied to the *Chorthippus parallelus* data in Table 1.2 if interpolation is used to match up the model stage frequencies with observed ones. A time unit of 5 days is then realistic for applying the equations given above. That is to say, the model can be used to generate population stage proportions at times 1, 2, . . ., 26, where these correspond to 5, 10, . . ., 130 days, respectively. The proportions for sample times 4, 9, . . ., 130 days can then be found using linear interpolation.

When this is done, the following estimates are found for population parameters, assuming that the 5-day survival rate is the same in each stage, and that the Weibull parameter α_j is also the same for each stage:

Parameter	Estimate	Standard error
Q_0	2.741	0.425
α_0	1.359	0.171
Q_1	2.663	0.445
Q_2	2.856	0.356
Q_3	1.348	0.424
Q_4	2.007	0.391
Common α	4.361	1.621
5-day survival	0.894	0.012

The estimated probabilities of entering stages 1–5 are 1.000, 0.724, 0.513, 0.422 and 0.326, from equation (4.39). Estimated distributions of stage durations are as follows:

	Probability of duration for stage			
Duration	1	2	3	4
0.0–1.0	0.014	0.010	0.238	0.047
–2.0	0.236	0.180	0.758	0.580
–3.0	0.565	0.520	0.004	0.370
–4.0	0.183	0.277	0.000	0.003
–5.0	0.003	0.013	0.000	0.000
–6.0	0.000	0.000	0.000	0.000
Mean	2.425	2.602	1.266	1.829
SD	0.693	0.735	0.432	0.566

These estimates are in reasonable agreement with those given in Example 4.1 from the KNM method. The KNM estimates of the mean stage durations in 5-day units are 1.72, 2.02, 1.64 and 1.98 for stages 1 to 4, respectively. The KNM estimate of the 5-day survival rate is 0.836.

The deviance of the fitted model is 87.18 with 54 degrees of freedom, which is significantly large at the 1% level. There is therefore evidence of extraneous variance over and above what is expected from Poisson sampling errors with the stage-frequencies. The heterogeneity factor is $H = 87.18/54 = 1.61$. A plot of observed and expected stage-frequencies (Figure 4.6) shows no pattern in the differences between these. An appropriate adjustment for the extraneous variance is therefore to simply multiply the standard errors in the above table by $\sqrt{1.61} = 1.27$.

When an attempt was made to fit the more general model for which survival rates vary from stage to stage this did not meet with much success. The MAXLIK algorithm could not increase the likelihood function to any appreciable extent and displayed all the characteristics of trying to fit an overparameterized model.

4.11   COMPARISON OF METHODS

Three methods for modelling of multi-cohort data have been discussed in detail in this chapter. In summary, they compare as follows:

1. The KNM method discussed in Sections 4.6 and 4.7 has the advantages of simplicity and of not requiring any particular assumptions about the form of the distribution of the time of entry to stage 1. Providing that the unit-time survival rate is constant, and sampling is continued long enough, this method has much to recommend it.
2. The Kempton model and its variations that are discussed in Sections 4.8 and 4.9 require the modelling of the distribution of the entry time to stage 1. This must be unimodal, at least. Given that the model assumed for data is approximately correct, this method should give about the most efficient means of estimating parameters since the full power of the method of maximum likelihood can be used to obtain these.
3. The Bellows and Birley model is more flexible than the other two. More parameters are involved, including separate survival parameters for each stage, although it is always possible to constrain some of these to be equal if overparameterization is a potential problem. The flexibility is obtained by using discrete time population changes. Care is needed to ensure that the time unit used is small enough for this to be a good approximation. This method is appropriate for cases where the KNM and Kempton models are inadequate to account for some aspects of the data – for example, where survival rates seem to vary from stage to stage.

EXERCISES

1. Table 2.5 shows Rigler and Cooley's (1974) estimates of the total numbers of individuals in six naupliar stages (N1–N6), five copepodite stages (CI–CV), and the adult stage for the copepod *Skistodiaptomus oregonensis* in Teapot Lake, Ontario, for the spring-to-autumn period in 1966. This is similar to other stage-frequency data considered in the present chapter, but there are more stages, and recruitment to stage 1 occurred for almost the entire sampling period. Also, as discussed in Example 2.5, the sampling variation seems to have been rather large.

    For this exercise, analyse the data by the KNM iterative method. The destruction of the population shortly after day 318 can be handled by

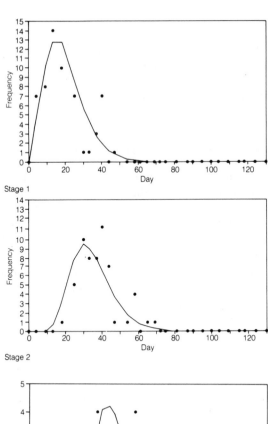

Stage 1

Stage 2

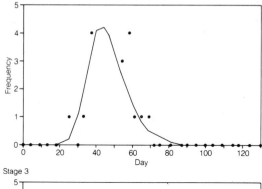

Stage 3

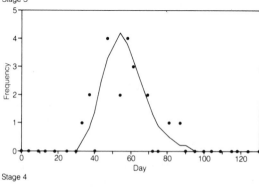

Stage 4

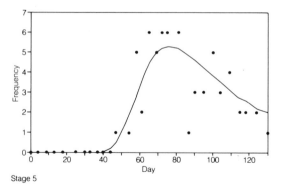

Stage 5

Figure 4.6 Comparison between observed and expected stage-frequencies for the Bellows and Birley model fitted to the grasshopper data of Table 1.2. Where ●, observed; —, expected.

regarding this as an example of sampling ending before all individuals are dead, on the assumption that the daily survival rate was more or less constant up to day 318. Type 4 simulation can be used to assess the accuracy of population estimates. Type 3 simulation can be used to assess the goodness of fit of the KNM model, although the excessive sampling variation suggests that the outcome will be the conclusion of a poor fit. Of course, the KNM model may be realistic with a poor fit simply being the result of sampling variation well in excess of what is expected with Poisson sampling errors. In that case, estimates will still be valid and standard errors can be approximately adjusted using an estimated heterogeneity factor. Such an adjustment will not, however, account for the apparent correlation between sampling errors for stage-frequencies at the same sample time. (See the comment at the end of Section 3.6.)

2. Analyse the *Orchesella cincta* data in Table 4.1, using physiological time in place of calendar time. Fit models assuming gamma, normal and inverse normal distributions for stage durations. In all cases, it is advisable to begin by fitting models with as many parameters as possible constrained to be equal. Models with more parameters can then be estimated by starting the maximum likelihood fitting procedure with the parameter values obtained from simpler models.

3. Analyse the *Orchesella cincta* data using Bellows and Birley's model, taking 1 day as the time unit. Take note of what is said in the previous exercise about fitting models in the order of the number of parameters.

5 Analysis of single cohort stage-frequency data

5.1 TYPES OF SINGLE COHORT DATA

This chapter is concerned with the analysis of stage-frequency data from populations where all individuals enter stage 1 at about the same time. Two types of data need to be considered. First, there are situations where the information is very similar in appearance to the multi-cohort data discussed in the previous chapter, but modelling is more straightforward because there is no need to account for an entry distribution. The total sample counts of stage-frequencies are expected to be proportional to the population size, which changes with time because of mortality. Second, there are situations where the total sample counts are not expected to be proportional to the population size, and all the information about development is contained in the sample proportions in different stages. It is possible to estimate distributions of stage durations, but no information is available about survival rates.

5.2 ANALYSIS USING MULTI-COHORT METHODS

In principle, any of the methods for analysing multi-cohort data that were discussed in the previous chapter can be used with the first type of single cohort data. The Kiritani–Nakasuji–Manly (KNM) method can be used more or less unchanged providing that the unit time survival rate is constant. The Kempton method and its variations has to be modified by removing the entry distribution to stage 1; a similar change is required with the Bellows and Birley model. Also, with single cohort data it is often the case that the number entering stage 1 (M_1) is known, and so does not require to be estimated.

The Bellows and Birley (1981) model is perhaps the most suitable of the three for single cohort data because of its greater flexibility. Removing the distribution of the entry time to stage 1 merely amounts to setting $P_1(0) = 1$ and $P_1(i) = 0, i > 0$ for equation (4.35). If M_1 is known then equation (4.41) is unnecessary.

Example 5.1 Analysis of a Single Cohort of *Callosobruchus chinensis*

The data shown in Table 5.1 were used by Bellows and Birley (1981) to illustrate their method of population estimation. Their paper should be consulted for a full description of how the data were obtained. Here, it suffices to note that a population of the bruchid pest *Callosobruchus chinensis* was allowed to develop in a large plastic box containing cowpeas. Some peas were taken from the box each day and dissected to see how many living individuals were present, and the stages that they were in. Because the number of hatched eggs could be counted on each day, the number of dead individuals was also known. A minimum of 50 live individuals were counted each day, and over the entire experiment an average of around 70 hatched eggs were counted each day. No mortality was considered in the egg stage.

One thing to note about these data is that they do not allow the distributions of stage durations to be estimated easily because of the small amount of

Table 5.1   Cohort stage-frequency data for the number of *Callosobruchus chinensis* per 100 hatched eggs. Stages I–IV are larvae, A1 is adults still in beans, A2 is adults after they have emerged from beans

Day	Eggs	I	II	III	IV	Pupae	A1	A2	Total
1	100.0								100.0
2	100.0								100.0
3	100.0								100.0
4	59.0	41.0							100.0
5		96.8							96.8
6		91.8							91.8
7		23.1	71.8						94.9
8		4.8	74.2	17.7					96.8
9		1.1	57.9	40.0					98.9
10		1.9	3.7	89.8	0.9				96.3
11			4.8	41.7	48.8				95.2
12			0.0	3.5	90.6				94.1
13			1.2	0.0	96.3				97.5
14				0.0	100.0				100.0
15				1.2	82.6	15.1			98.8
16					37.7	58.0			95.7
17					10.6	74.2			84.9
18					10.5	81.4		1.2	93.0
19					2.1	70.8	18.8	0.0	91.7
20						22.6	72.6	1.2	96.4
21						8.2	75.3	16.4	100.0
22						4.8	38.6	53.0	96.4
23							12.7	85.9	98.6
24							8.5	91.5	100.0
25							2.0	88.2	90.2
26								88.4	88.4

Table 5.2 Estimates for the *Callosobruchus chinensis* data obtained from the iterative KNM method

Stage	Stage-specific survival rate	Duration	Number entering stage
1 Eggs	1.000	3.1*	100.0
2 Instar I	0.972	2.3	100.0
3 Instar II	0.976	2.0	97.2
4 Instar III	0.978	1.8	94.9
5 Instar IV	0.944	4.7	92.7
6 Pupae	0.958	3.4	87.5
7 Adults in pea	0.976	2.0	83.9
8 Emerged adults	–	–	81.9

*Estimated directly from the data (see text).

variation in some of these durations. This is evident because the number of days during which stages are present does not vary much after stage 1. Thus, stages 2–7 are present for 7, 7, 8, 10, 8 and 7 days, respectively. If stage durations are very variable we would expect to see successive stages covering longer and longer periods. Here, if anything, there seems to be a reduction in variation from stage 5 onwards.

Because stage durations were apparently rather constant it is interesting to analyse the data initially with the KNM model. No mortality was considered in the egg stage, and mortality rates were clearly very low in the other stages. It is therefore reasonable to assume a constant daily survival rate in stages 2–8, and apply the estimation equations of Sections 4.6 and 4.7 to these stages only. This then allows the estimation of the durations and stage-specific survival rates of stages 2–7, and a daily survival rate for stages 2–8. As the experiment ended before all individuals were dead, the iterative method of Section 4.7 has to be used. This converges satisfactorily with the data for stages 2–8, and provides the estimates shown in Table 5.2.

The mean duration of the egg stage can be estimated directly from the stage-frequencies given in Table 5.1. It can be seen that 41% of sampled eggs had a stage duration between 2 and 3 days (say 2.5 days) and 59% a duration between 3 and 4 days (say 3.5 days), which gives a mean duration of 3.1 days.

The numbers entering stages given in the table were determined by applying the stage-specific survival rates to the known number (100%) entering instar I. For stages 7 and 8, the estimates seem to be too low in comparison with the data stage-frequencies, which suggest that around 90% entered stage 8. However, this discrepancy between the estimates and the data is not surprising given the low total frequencies for some of the samples in the middle of the sampling period.

An attempt was made to fit three versions of the Bellows and Birley model to the data. For estimation purposes, frequencies were multiplied by 10 and treated as integer counts in order to apply the MAXLIK algorithm for finding maximum likelihood estimates. As sample sizes averaged 70 per day, the data 'counts' are more like counts multiplied by $1000/70 = 14.29$. However, this only becomes an important consideration when it comes to questions of the accuracy of estimates. As noted in Section 3.6, multiplying the standard errors output by the MAXLIK algorithm by the square root of the heterogeneity factor will make an allowance for this.

Table 5.3 Estimates obtained from fitting the Bellows and Birley model with constant α values for stage durations and a constant daily survival rate. The standard error values are adjusted for extraneous variance (see text)

Parameter	Estimate	Standard error
q_1	3.373	0.139
q_2	2.543	0.229
q_3	1.730	0.229
q_4	1.687	0.303
q_5	4.750	0.287
q_6	3.312	0.336
q_7	2.059	0.354
α	6.568	0.500
ϕ	0.998	0.066

The first model considered was one with the Weibull distributions in stages having the same α value, but different Q values (equation (4.37)). Also, the daily survival rate was assumed to be the same in all stages. This gave nine parameters to be estimated. The values obtained are shown in Table 5.3. The second model considered was the same as the first but with the α value allowed to vary from stage to stage. Convergence problems were encountered in this case with a clear indication of overparameterization. The third model considered had a constant α value, but daily survival rates were allowed to vary from stage to stage. For this model, the MAXLIK algorithm converged. However, several of the estimated survival rates were larger than unity, and the deviance of the fitted model was hardly changed from the value from the first model. Consequently, the first model was accepted as the most reasonable.

The heterogeneity factor for this model is 67.21, which is much higher than the 14.29 expected from the ratio of the stage-frequencies used in the estimation to the original counts of live individuals. Apparently, there was a considerable amount of extraneous variance. The standard errors shown in

Table 5.3 are adjusted for this extraneous variance, being the values output by MAXLIK multiplied by $\sqrt{67.21} = 8.20$.

The distributions estimated for stage durations are shown in the table that follows. The means agree reasonably well with the KNM estimates shown in Table 5.2.

Days	Duration of stage						
	1	2	3	4	5	6	7
0.0–1.0	0.000	0.002	0.027	0.032	0.000	0.000	0.009
–2.0	0.031	0.185	0.898	0.921	0.003	0.035	0.554
–3.0	0.339	0.762	0.075	0.047	0.044	0.371	0.437
–4.0	0.583	0.052	0.000	0.000	0.229	0.561	0.000
–5.0	0.047	0.000	0.000	0.000	0.477	0.032	0.000
–6.0	0.000	0.000	0.000	0.000	0.237	0.000	0.000
–7.0	0.000	0.000	0.000	0.000	0.010	0.000	0.000
Mean	3.144	2.363	1.548	1.515	4.429	3.088	1.929
SD	0.623	0.475	0.315	0.280	0.841	0.615	0.512

Stage-specific survival rates are estimated as 0.992, 0.995, 0.995, 0.996, 0.989, 0.993 and 0.994 for stages 1–7, respectively. The estimated proportions entering stages 1–8 are 1.000, 0.992, 0.987, 0.982, 0.978, 0.968, 0.961 and 0.955. These seem quite realistic from the data. Figure 5.1 shows a graphical comparison of observed and expected frequencies. From this point of view the model seems a reasonable fit.

5.3 DATA WITHOUT MORTALITY

Now consider situations where the total stage-frequencies at different sample times are not expected to be proportional to the population size, and are best thought of as being arbitrary. The method of Kiritani–Nakasuji–Manly then breaks down completely. However, the other models discussed in the previous chapter can still be used to analyse data by setting survival parameters to whatever is appropriate for no deaths, and making the total observed and expected frequencies agree for each sample. There are, however, some alternative approaches that either involve simpler calculations, or fit in better with methods used in the mainstream of statistical analysis. These will be discussed in the remainder of this chapter.

Two types of situation are covered by the methods that will now be considered. First, cases arise where mortality either does not occur or is negligible. The population size is then constant for all sample times and it is the proportions in different stages at different times that can be used to estimate distributions of stage durations. The total stage-frequencies at different sample times only provide information on the accuracy of estimates. Second, there are cases where mortality does occur, but for some reason it is not possible to collect data with the total stage-frequency at each sample time expected to be proportional to the population size. For example, weather conditions may cause fluctuations in sample size that are so large that they completely swamp the small amount of variation due to the changing population size. Again, total stage-frequencies at different sample sizes are best thought of as being arbitrary, and giving information only about the accuracy of estimates.

An example of the first type of situation is provided by the data in Table 1.3 on the breast development of New Zealand schoolgirls. These results were not obtained by sampling a single cohort of girls over a period of 5 years. Rather, they were obtained by doing a survey over a fairly short period of time of girls that were born from 10 to 15 years earlier. However, it is reasonable to regard these data as being equivalent to what would be obtained by sampling all the girls born in one year when they are aged from 10 to 11 years, again when they are aged from 11 to 12, and so on, because there is no reason to believe that rates of breast development change appreciably for girls born within 5 years of each other. Mortality rates in New Zealand girls aged from 10 to 15 years are obviously very low and can be ignored when studying breast development. The increasing number of girls sampled as age increased is probably merely the result of the survey process tending to favour older girls. It certainly does not reflect the numbers of girls of different ages in the population.

The second type of situation, where mortality occurs but there is no information about it, can be treated in exactly the same way as the first type of situation. However, the way that results are interpreted may depend on whether or not it is reasonable to assume that the mortality rate is the same in all stages. If the mortality rate is the same in all stages then the survivors in the population at any time are a random sample from the population before mortality. A random sample from the population of survivors is then a random sample from a random sample, which is equivalent to a random sample taken from the population that would have existed without any deaths. Hence, the parameters estimated for stage durations are the ones that would apply to the population without deaths.

It is a different matter if the mortality rate varies from stage to stage. Then the individuals that spend a long time in a stage with high mortality are less likely to survive than the individuals that pass quickly through the stage. Survival and development are no longer independent and the parameters estimated for stage durations are ones that apply for the population after deaths. The distinction between the stage durations that would occur without mortality and the stage durations with mortality may or may not be important in practice, but it should be appreciated that it exists (Section 4.3).

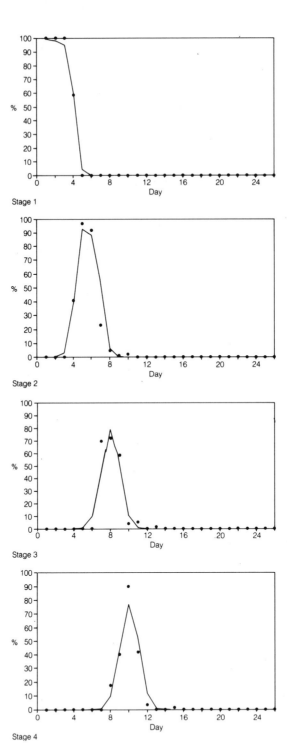

Stage 1

Stage 2

Stage 3

Stage 4

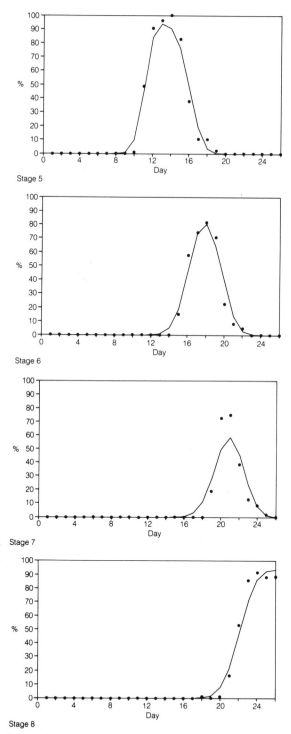

Figure 5.1 Comparison of observed and expected stage-frequencies for the Bellows and Birley model fitted to the *Callosobruchus chinensis* data of Table 5.1. Where ●, observed; —, expected.

5.4   NON-PARAMETRIC ESTIMATION

Pontius *et al.* (1989a,b) proposed a simple non-parametric approach to the estimation of the mean times required to reach stages when samples are taken over the period from when a cohort starts developing until all individuals are in the last stage. This is based on the result that the mean time to reach stage j is given by

$$\mu_j = \int_0^\infty x g_j(x)\,dx = \int_0^\infty G_j(x)\,dx,$$

where $g_j(x)$ is the probability density function of the time of entry to stage j, and $G_j(x)$ is the probability that the time of entry to stage j is after x (i.e., the integral of $g_j(x)$ from x to infinity). The above integral for $G_j(x)$ can be approximated using the trapezoidal rule to give

$$\mu_j \approx \frac{1}{2} \sum_{i=0}^{n-1} \{G_j(t_i) + G_j(t_{i+1})\}\{t_{i+1} - t_i\}.$$

Let samples be taken at times $t_1, t_2, \ldots, t_n$, where all individuals are in stage 1 at time $t_0 = 0$ (which is before time t_1), and all individuals are in stage q at time t_n. Also, let p_{ij} be the proportion of individuals in the sample at time t_i that are not yet in stage i, with $p_{i1} = 0$, for all i. Then, p_{ij} is an estimate of $G_j(t_i)$ so an estimate of μ_j is given by

$$m_j = \frac{1}{2} \sum_{i=0}^{n-1} (p_{ij} + p_{i+1j})(t_{i+1} - t_i),$$

which can also be written as

$$m_j = \frac{1}{2} t_1 + \frac{1}{2} \sum_{i=1}^{n-1} p_{ij}(t_{i+1} - t_{i-1}). \tag{5.1}$$

Pontius *et al.* note that if the sample proportions p_{ij} are independent for different sample times, and are binomially distributed then it follows that the variance of m_j can be estimated by

$$\text{var}(m_j) = \frac{1}{4} \sum_{i=1}^{n-1} p_{ij}(1 - p_{ij})(t_{i+1} - t_{i-1})^2/n_i, \tag{5.2}$$

where n_i is the total sample frequency at time t_i.

Two equations for estimating the variance of the time required to reach stage j are given by Pontius *et al.* The simplest is

$$V_j = \frac{1}{2} \sum_{i=0}^{n-1} (p_{ij} + p_{i+1j})(t_{i+1}^2 - t_i^2) - \mu_j^2. \tag{5.3}$$

Actually, what is usually of interest is the means and variances of the durations of stages. The duration of stage j can be estimated by

$$a_j = m_{j+1} - m_j. \tag{5.4}$$

Assuming that stage durations are distributed independently for the same individual, the variance of the duration of stage j can be estimated by

$$s_j^2 = V_{j+1} - V_j. \tag{5.5}$$

Pontius *et al.* provide the following equation for the variance of an estimated stage duration:

$$\text{var}(a_j) = \frac{1}{4} \sum_{i=1}^{n-1} (p_{ij+1} - p_{ij})(1 - p_{ij+1} + p_{ij})(t_{i+1} - t_{i-1})^2 / n_i. \tag{5.6}$$

This is based on the assumption that the proportions p_{ij} are binomially distributed. If there is more variation than the binomial distribution suggests then the true variance of a_j will be larger than what is given by this equation.

Example 5.2   Estimation of Stage Durations for *Callosobruchus chinensis*

The data in Table 5.1 from a laboratory population of *Callosobruchus chinensis* were analysed by Bellows and Birley's model in Example 5.1. It is also suitable for the method of Pontius *et al.* (1989a,b). Table 5.4 shows the values for the sample proportions that are required for equations (5.1) to (5.6). The estimated stage durations and their standard errors are as follows:

		Stage					
	1	2	3	4	5	6	7
a_j	3.09	2.73	2.21	2.01	4.96	3.65	2.35
Standard error	0.06	0.09	0.10	0.10	0.12	0.13	0.12

In calculating the standard errors, sample sizes of $n_i = 70$ were used, this being the average number sampled per day. As noted above, these standard errors will be too small if the sample stage-frequencies are more variable than is expected from binomial proportions. The calculations in Example 5.1 suggest that this is indeed the case.

There is fairly reasonable agreement between these estimates of mean stage durations and those calculated in Example 5.1 using the Bellows and Birley model (3.14, 2.36, 1.55, 1.52, 4.43, 3.09 and 1.93, for stages 1–7,

Table 5.4 Values of p_{ij} needed for the estimation of mean stage durations by the method of Pontius *et al.* (1989a,b)

i	p_{i1}	p_{i2}	p_{i3}	p_{i4}	p_{i5}	p_{i6}	p_{i7}	p_{i8}
1	0.00	1.00	1.00	1.00	1.00	1.00	1.00	1.00
2	0.00	1.00	1.00	1.00	1.00	1.00	1.00	1.00
3	0.00	1.00	1.00	1.00	1.00	1.00	1.00	1.00
4	0.00	0.59	1.00	1.00	1.00	1.00	1.00	1.00
5	0.00	0.00	1.00	1.00	1.00	1.00	1.00	1.00
6	0.00	0.00	1.00	1.00	1.00	1.00	1.00	1.00
7	0.00	0.00	0.24	1.00	1.00	1.00	1.00	1.00
8	0.00	0.00	0.05	0.82	1.00	1.00	1.00	1.00
9	0.00	0.00	0.01	0.60	1.00	1.00	1.00	1.00
10	0.00	0.00	0.02	0.06	0.99	1.00	1.00	1.00
11	0.00	0.00	0.00	0.05	0.49	1.00	1.00	1.00
12	0.00	0.00	0.00	0.00	0.04	1.00	1.00	1.00
13	0.00	0.00	0.00	0.01	0.01	1.00	1.00	1.00
14	0.00	0.00	0.00	0.00	0.00	1.00	1.00	1.00
15	0.00	0.00	0.00	0.00	0.01	0.85	1.00	1.00
16	0.00	0.00	0.00	0.00	0.00	0.39	1.00	1.00
17	0.00	0.00	0.00	0.00	0.00	0.13	1.00	1.00
18	0.00	0.00	0.00	0.00	0.00	0.11	0.99	0.99
19	0.00	0.00	0.00	0.00	0.00	0.02	0.79	1.00
20	0.00	0.00	0.00	0.00	0.00	0.00	0.23	0.99
21	0.00	0.00	0.00	0.00	0.00	0.00	0.08	0.84
22	0.00	0.00	0.00	0.00	0.00	0.00	0.05	0.45
23	0.00	0.00	0.00	0.00	0.00	0.00	0.00	0.13
24	0.00	0.00	0.00	0.00	0.00	0.00	0.00	0.09
25	0.00	0.00	0.00	0.00	0.00	0.00	0.00	0.02
26	0.00	0.00	0.00	0.00	0.00	0.00	0.00	0.00

respectively). However, it is probably no accident that the Bellows and Birley estimates are lower than the Pontius *et al.* ones, except in stage 1.

Unfortunately, equation (5.5) yields some negative estimates for the variances of stage durations as the estimated variances for times of entering stages do not always increase with the stage number. This may be a common problem with real data sets.

5.5 PARAMETRIC MODELS FOR ESTIMATION

A parametric model for stage-frequency data without mortality is one providing an equation for the proportion in stage j at time t as a function of one or more parameters that have to be estimated from data. In practice, the choice of the form of the equation will often be arbitrary. However, as this function is only really being used to smooth the data, the choice made may not be crucial, with most sensible functions giving rather similar results. Two models that are plausible are the multiple logistic, and what can be called a log–log model. These are conveniently defined in terms of $Q_j(t)$, the probability that an individual exceeds stage j at time t. Multiple logistic models take the form

$$Q_j(t) = \exp\{h_j(t)\}/[1 + \exp\{h_j(t)\}],$$

where $h_j(t)$ is an increasing function of t. McCullagh (1983) suggested that at time t_i, $h_j(t_i)$ might take one of the forms $h_j(t_i) = m_j + nt_i$, $h_j(t_i) = m_j + r_i$, or $h_j(t_i) = m_j + n_j t_i$. The description logistic is used here because the logistic transformation of $Q_j(t)$, which is $\log[Q_j(t)/\{1 - Q_j(t)\}]$, is equal to $h_j(t)$.

For the log–log model $- \log[- \log\{Q_j(t)\}] = h_j(t)$, where again $h_j(t)$ is an increasing function of the time t. The inverse relationship is then $Q_j(t) = \exp[- \exp\{ - h_j(t)\}]$. McCullagh (1983) suggested taking $h_j(t)$ as $m_j + n_j(t - 10)$ for the data on breast development of New Zealand schoolgirls (Table 1.3), and found the following estimated parameters by maximum likelihood: $h_1(t) = -0.594 + 1.016(t - 10)$, $h_2(t) = -1.455 + 0.840(t - 10)$, $h_3(t) = -1.895 + 0.608(t - 10)$, and $h_4(t) = -2.223 + 0.420(t - 10)$. The chi-squared goodness-of-fit statistic for this model is 25.4 with 12 degrees of freedom. This is significantly large at the 5% level but this may reflect only extraneous variance. To fit a model like this by maximum likelihood, it should be noted that the probability of being in stage j at time t is $p_j(t) = Q_{j-1}(t) - Q_j(t)$. The methods of estimation discussed in Chapter 3 can then be used.

A multiple logistic type of model was suggested by Dennis et al. (1986) to model insect population development. This followed an approach used by Osawa et al. (1983) where the cumulative density function of the normal distribution takes the place of the logistic. Rather earlier, Aitchison and Silvey (1957) also estimated mean stage durations using a normal distribution model ignoring mortality.

5.6   ESTIMATING THE DURATIONS OF STAGES

One limitation with fitting a parametric model as just discussed is that it provides no immediate information on the durations of stages because there is no direct relationship between the estimated parameters and durations. Indeed, the value of fitting this type of model can be questioned, because all it does is 'smooth' the estimated proportions in stages by fitting an equation to them.

However, it is possible to derive estimates of the means and standard deviations of stage durations using the Pontius et al. (1989a,b) equations given above, providing that it is acceptable to extrapolate the fitted parametric model to determine the numbers in stages outside the range of the data. The following example indicates how this can be done.

Example 5.3   Estimating Stage Durations for Breast Development of New Zealand Schoolgirls

As noted above, when a log–log model is fitted to the data on the breast development of New Zealand schoolgirls (Table 1.3), the proportions above the stages 1–4 are estimated to be given by the following equations:

$$Q_1(t) = \exp[-\exp\{-0.594 + 1.016(t-10)\}],$$

$$Q_2(t) = \exp[-\exp\{-1.455 + 0.840(t-10)\}],$$

$$Q_3(t) = \exp[-\exp\{-1.895 + 0.608(t-10)\}],$$

and

$$Q_4(t) = \exp[-\exp\{-2.223 + 0.420(t-10)\}],$$

These give the percentages in different stages at different ages that are shown in Table 5.5, where, for example, girls aged 10–11 years are assigned the age 10.5.

Table 5.5 Percentages of New Zealand schoolgirls with different stages of breast development according to a log–log model fitted to the data in Table 1.3

| | *Percentages in stage* | | | | |
Age	1	2	3	4	5
6.5	100.0	0.0	0.0	0.0	0.0
7.5	100.0	0.0	0.0	0.0	0.0
8.5	100.0	0.0	0.0	0.0	0.0
9.5	95.1	4.8	0.1	0.0	0.0
10.5	66.4	27.6	5.3	0.7	0.1
11.5	32.6	37.7	22.8	6.2	0.7
12.5	13.3	27.5	35.8	19.4	3.9
13.5	5.0	15.2	34.4	33.3	12.0
14.5	1.9	7.5	25.7	40.2	24.8
15.5	0.7	3.5	16.8	39.1	40.0
16.5	0.2	1.6	10.2	33.2	54.8
17.5	0.1	0.7	5.9	26.0	67.3
18.5	0.0	0.3	3.4	19.2	77.1
19.5	0.0	0.1	1.9	13.7	84.3
20.5	0.0	0.1	1.1	9.5	89.4
21.5	0.0	0.0	0.6	6.5	92.9
22.5	0.0	0.0	0.3	4.4	95.3
23.5	0.0	0.0	0.2	3.0	96.9
24.5	0.0	0.0	0.1	2.0	97.9
25.5	0.0	0.0	0.1	1.3	98.6

If the Pontius *et al.* equations are applied to these percentages, assuming that development starts at age 6.5, then the mean ages of entry to stages 1–5 are estimated at 6.5, 11.2, 12.4, 14.1 and 16.7 years, respectively, with standard deviations of 0, 1.4, 1.6, 2.2 and 3.1 years, respectively. Taking differences between the mean entry times then gives estimated durations for stages 1–4 of 4.7, 1.3, 1.7 and 2.6 years, respectively. From equation (5.5) the standard deviations of the durations of these stages are estimated at 1.4, 0.9, 1.5 and 2.1 years, respectively.

The estimated mean times of entry to stages and mean stage durations seem realistic. However, the extrapolated percentages in stages for the higher ages shown in Table 5.5 do not seem very realistic as presumably almost all girls would be fully developed by age 21. It is likely, therefore, that the spread of development times for stages 1–4 have been somewhat exaggerated and that as a result of this the standard deviations of times to enter stages and stage durations are somewhat overestimated.

EXERCISES

1. Table 5.6 shows a set of artificial data that was generated on a computer using the Bellows and Birley model. This was obtained by letting 1000 individuals enter stage 1 just before time 1. They were then subjected to unit-time survival rates of 0.8, 0.7, 0.9, 0.8 and 0.8 in stages 1–5, respectively. The distributions of stage durations were as follows:

	Probabilities for stages			
Duration	*1*	*2*	*3*	*4*
0–1	0.25	0.25	0.00	0.00
–2	0.50	0.50	0.25	0.25
–3	0.25	0.25	0.50	0.50
–4	0.00	0.00	0.25	0.25
Mean	1.50	1.50	2.50	2.50
SD	0.71	0.71	0.71	0.71

Analyse the data using the Bellows and Birley model to see how well the true values for the population parameters and the stage distributions can be recovered.

2. Estimate the means and standard deviations of stage durations for the artificial set of data given in Table 5.6, using the Pontius *et al.* (1989a,b) equations. See how these compare with the values used in generating the data, and the estimates obtained from the Bellows and Birley model.

Table 5.6   Simulated stage-frequency data for a single cohort of 1000 individuals

	Stage				
Time	*1*	*2*	*3*	*4*	*5*
1	1003	0	0	0	0
2	637	200	0	0	0
3	163	441	38	0	0
4	0	309	155	0	0
5	0	109	221	8	0
6	0	19	225	58	0
7	0	0	140	104	0
8	0	0	78	157	3
9	0	0	20	134	31
10	0	0	3	88	51
11	0	0	0	47	68
12	0	0	0	10	81
13	0	0	0	4	66
14	0	0	0	0	65
15	0	0	0	0	31
16	0	0	0	0	39
17	0	0	0	0	33
18	0	0	0	0	18
19	0	0	0	0	22
20	0	0	0	0	15
21	0	0	0	0	17
22	0	0	0	0	16
23	0	0	0	0	7
24	0	0	0	0	8
25	0	0	0	0	5
26	0	0	0	0	6
27	0	0	0	0	6
28	0	0	0	0	4
29	0	0	0	0	3
30	0	0	0	0	4

3. Consider again the computer-generated data in Table 5.6. Fit logistic and log–log model to the data for sample times 2–10 only. Use whichever model fits best to estimate stage-frequencies before and after these sample times and then use the Pontius *et al.* equations to estimate the means and standard deviations of stage durations. The estimates can be compared with those obtained from the full data in Exercise 2, and also the true values that are provided in Exercise 1.

6 Matrix and other models for reproducing populations

6.1 CONTINUALLY REPRODUCING POPULATIONS

The important characteristic of the situations that will be considered in this chapter is that the population being studied is continually reproducing itself. What is of interest is to find a model that makes it possible to determine the future structure of the population from a knowledge of the present structure. If this is too optimistic, just gaining some understanding of how numbers in different stages at one point in time determine numbers one unit of time later may be of some value.

An example set of data is shown in Table 1.4. This was recorded by Lefkovitch (1964a) for a laboratory population of the cigarette beetle *Lasioderma serricorne* started with 80 adults. In this case, the individuals develop through the stages eggs → larvae → pupae → adults, with losses through deaths in all stages. The adults then produce eggs to continue the cycle.

Most of the mathematical models that have been developed to account for the dynamics of reproducing populations assume that the ages of individuals are known. When this is true, there are basically two alternative approaches that have been used. The first is the continuous-time integral equation method pioneered by Sharpe and Lotka (1911). The second uses grouped age intervals and a matrix formulation, and was first proposed independently by Bernardelli (1941), Lewis (1942) and Leslie (1945). Both of these approaches are reviewed by Pollard (1973).

In the context of this book it is the matrix approach that is most relevant because several people have developed the ideas of Bernardelli, Lewis and Leslie to allow for stage rather than age grouping of individuals. The next section of this chapter therefore contains a brief description of the original matrix model for age-grouped populations and is followed by sections on the extensions that have been suggested for analysing stage-grouped populations. The final section of the chapter concerns methods for estimating population parameters using the ratio of egg to adult numbers.

6.2 THE BERNARDELLI–LESLIE–LEWIS MODEL

The matrix approach to modelling the dynamics of a population is usually attributed to Leslie (1945, 1948), although, as mentioned above, the same model was proposed earlier by Bernardelli (1941) and Lewis (1942). It involves considering a population at discrete points in time $0, 1, 2, \ldots$, with individuals in age groups $0, 1, 2, \ldots, k$. Age group x comprises all individuals with an age from x until just less than $x + 1$. For convenience, only females are counted.

The following notation will be used in the present discussion:

$n(x, t)$ = the number of females in the age group x at time t;
$p(x)$ = the probability that a female in the age group x at time t will survive to be in the age group $x + 1$ at time $t + 1$; and
$B(x)$ = the average number of female offspring born to females aged from x to $x + 1$ in a unit period of time that survive to the end of that period.

The number of females in age group 0 at time $t + 1$ will then be the sum of the offspring from females of different ages:

$$n(0, t+1) = B(0)n(0, t) + B(1)n(1, t) + \ldots + B(k)n(k, t) \tag{6.1}$$

It also follows from the definitions that

$$n(x+1, t+1) = p(x)n(x, t) \tag{6.2}$$

for $x = 0, 1, \ldots, k - 1$.

Equations (6.1) and (6.2) can be written together as the matrix equation

$$\begin{bmatrix} n(0, t+1) \\ n(1, t+1) \\ \cdot \\ \cdot \\ \cdot \\ n(k, t+1) \end{bmatrix} = \begin{bmatrix} B(0) & B(1) & \cdots & B(k-1) & B(k) \\ p(0) & 0 & \cdots & & 0 \\ \cdot & & & & \\ \cdot & & & & \\ \cdot & & & & \\ 0 & 0 & \cdots & p(k-1) & 0 \end{bmatrix} \begin{bmatrix} n(0, t) \\ n(1, t) \\ \cdot \\ \cdot \\ \cdot \\ n(k, t) \end{bmatrix}$$

or

$$\mathbf{N}_{t+1} = \mathbf{M}\mathbf{N}_t. \tag{6.3}$$

It then follows that

$$\mathbf{N}_t = \mathbf{M}^t \mathbf{N}_0. \tag{6.4}$$

The matrix $\mathbf{M}$, whose elements are the fecundity rates $B(x)$ and the survival probabilities $p(x)$, is usually called the Leslie matrix. Equation (6.4) shows that the numbers in different age groups at an arbitrary time t are determined by the numbers in the age groups at time zero ($\mathbf{N}_0$) and the Leslie matrix raised to the power t. Subject to certain mild assumptions it is possible to show that a population following this model will eventually reach a stable distribution for the relative numbers of individuals with different ages, and be growing or declining at a constant rate. The long-term behaviour of the population is determined by the dominant eigenvalue of the Leslie matrix. (See Pollard (1973, ch. 4) for a discussion of the various theoretical results that are known.)

Plant (1986) considered the question of how to determine the elements of the transition matrix when the object is to model a population with a relatively small number of age groups. This may be important, for example, with simulation studies where a large transition matrix may be impractical because of the computing times involved. Plant provided formulae for survival probabilities and fecundity rates for grouped ages and worked a numerical example involving the modelling of the infestation of the Mediterranean fruit fly *Ceratitis capitata* in California.

6.3   LEFKOVITCH'S MODEL FOR POPULATIONS GROUPED BY LIFE STAGES

Lefkovitch (1963, 1964a, 1964b, 1965) modified the Bernardelli–Leslie–Lewis model to allow a population to be grouped by life stages rather than by age. He did this by the simple expedient of allowing the number in stage j at time $t + 1$ to depend on the numbers in all previous stages at time t. Thus, if $f_j(t)$ is the number of individuals in stage j at time t then for q stages his model is, in matrix notation,

$$
\begin{bmatrix}
f_1(t+1) \\
f_2(t+1) \\
\cdot \\
\cdot \\
\cdot \\
f_q(t+1)
\end{bmatrix}
=
\begin{bmatrix}
m_{11} & m_{12} & \cdots & m_{1q} \\
m_{21} & m_{22} & \cdots & m_{2q} \\
\cdot \\
\cdot \\
\cdot \\
m_{q1} & m_{q2} & \cdots & m_{qq}
\end{bmatrix}
\begin{bmatrix}
f_1(t) \\
f_2(t) \\
\cdot \\
\cdot \\
\cdot \\
f_q(t)
\end{bmatrix}. \quad (6.5)
$$

or

$$
\mathbf{F}_{t+1} = \mathbf{M}\mathbf{F}_t,
$$

so that

$$
\mathbf{F}_t = \mathbf{M}^t \mathbf{F}_0. \quad (6.6)
$$

The typical entry in the matrix $\mathbf{M}$ in equation (6.5), m_{ij}, reflects how the number in stage i at time $t + 1$ depends on the number in stage j at time t. Equation (6.6) is similar to equation (6.4). However, the matrix $\mathbf{M}$ of the latter equation does not have the simple structure of a Leslie matrix, with its many necessarily zero elements. There is an implicit assumption with Lefkovitch's model that the age distribution within stages is constant enough to make any variation in the m_{ij} values with time unimportant.

Lefkovitch's model is not as straightforward as the Bernardelli–Leslie–

Lewis model to study from a theoretical point of view. Nevertheless, as noted by Lefkovitch, the long-term behaviour of the population will be determined by the eigenvalue of **M** with maximum modulus, with its corresponding eigenvector. To be precise, let τ_1 be the eigenvalue of the matrix **M** with the largest absolute value, with corresponding eigenvector v_1. Then, the proportions in different stages in the population should tend towards the proportions in the vector v_1, and the numbers in each stage should increase by the factor τ_1 per generation. If the dominant eigenvalue of the matrix **M** is complex then cyclic population changes are implied. In this case, it may be that the real situation is that the m_{ij} values were not constant over the sampling period. A negative dominant eigenvalue is biologically meaningless.

Although eigenvalues and vectors can be used to predict population behaviour, a more empirical approach may be more useful. Once **M** is estimated it can be used to simulate populations to determine likely patterns of behaviour. This approach makes it possible to introduce random perturbations between sample times and make the model closer to reality. As will be discussed below, the behaviour of a model with added random variation can be very different from what is expected from the model without random variation.

If the numbers in the different stages are known for a series of at least q^2 equally spaced sample times then the constants in the matrix **M** can be determined by regression. For stage j alone, equation (6.5) gives

$$f_j(t+1) = m_{j1}f_1(t) + m_{j2}f_2(t) + \ldots + m_{jq}f_q(t). \tag{6.7}$$

Estimates of $m_{j1}, m_{j2}, \ldots, m_{jq}$ can be found, therefore, by a multiple regression of $f_j(t+1)$ values on the numbers in the different stages at time t. The regression should not include a constant term. Combining the regression estimates of m values for different stages leads to the estimate of **M** suggested by Lefkovitch, although this is not immediately obvious.

There are practical difficulties with using this multiple regression method to estimate **M**. The number of parameters to be estimated may be rather large for available data so that estimates are subject to large sampling errors. This may then mean that estimates are not biologically meaningful. In particular, negative values will be difficult to interpret because they imply that negative stage-frequencies can occur. Lefkovitch suggested overcoming this problem by (1) setting m_{ij} values to zero rather than estimating them whenever biological knowledge allows this; (2) setting negative estimates of m_{ij} to zero, and re-estimating the remaining parameters; and (3) replacing estimates that are not significantly different from zero with zero, and re-estimating the remaining parameters. There is a problem with (3) as the usual assumptions of multiple regression will not apply with equation (6.7). Hence, it is not valid to use this theory to determine the standard errors of estimated regression coefficients.

Another problem is the natural tendency for the multiple regression estimates to be ill-conditioned. If a population follows the Lefkovitch model exactly, and has been developing long enough for stage-frequencies to be determined largely by the eigenvalue of **M** with maximum modulus, then stage-frequencies in successive stages will be more or less proportional. This will make the inverse matrix in the multiple regression almost singular. Thus, good estimates of the elements of **M** can be expected from data only on populations that have been started for a short time so that the relative frequencies in different stages are still changing substantially. Unfortunately, this is just the situation where the m_{ij} values are liable to be changing because of changes in age distributions within stages.

A question that sometimes arises is the determination of appropriate boundaries for the stages in a population when these depend on some continuous character such as size. If too few stages are used then the elements in the transition matrix will vary too much with the age structure within stages. If too many stages are used then it will be difficult to get good estimates of the large number of parameters involved. Vandermeer (1978) proposed an algorithm for getting an appropriate balance between these two extremes, which was later revised and extended by Moloney (1986).

6.4 USHER'S MODEL

The main problem with using Lefkovitch's model is the large number of coefficients to be estimated in the transition matrix. However, this can be overcome to some extent if the time between samples is small so that the possibility of an individual developing through more than one stage in this time can be discounted. A model based on this assumption was developed by Usher (1966, 1969) in the context of the management of a forest. The population then consists of trees in developmental stages that are determined by their size. However, the model can be applied with other definitions for stages. Usher assumed that equation (6.5) can be simplified to

$$
\begin{bmatrix}
f_1(t+1) \\
f_2(t+1) \\
f_3(t+1) \\
\cdot \\
\cdot \\
\cdot \\
f_{q-1}(t+1) \\
f_q(t+1)
\end{bmatrix}
=
\begin{bmatrix}
B_1 & B_2 & B_3 & \ldots & B_{q-1} & B_q \\
b_1 & a_2 & 0 & \ldots & 0 & 0 \\
0 & b_2 & a_3 & \ldots & 0 & 0 \\
\cdot & & & & & \\
\cdot & & & & & \\
0 & 0 & 0 & \ldots & a_{q-1} & 0 \\
0 & 0 & 0 & \ldots & b_{q-1} & a_q
\end{bmatrix}
\begin{bmatrix}
f_1(t) \\
f_2(t) \\
f_3(t) \\
\cdot \\
\cdot \\
\cdot \\
f_{q-1}(t) \\
f_q(t)
\end{bmatrix}
\tag{6.8}
$$

Here, B_j is the contribution to the number in stage 1 at time $t+1$ that comes from those in stage j at time t. For $j = 1$, this contribution comes from those that remain alive but do not develop to stage 2, and also possibly from the reproduction of stage 1 individuals. For $j > 1$, the contribution is from reproduction only. Also, a_j is the probability that an individual in stage i at time t is still in stage j at time $t+1$, while b_j is the probability that an individual in stage j at time t moves to stage $j+1$ by time $t+1$. The sum $a_j + b_j$ gives the survival rate between two sample times for an individual in stage j at the first of these times.

As for Lefkovitch's model, the dominant eigenvalue of the transition matrix indicates the long-term behaviour of the population and the corresponding eigenvector indicates the stable proportions in different stages. If the dominant eigenvalue is τ_1, which is real, then the population size will be multiplied by τ_1 each time unit once these stable proportions have been reached.

In principle, the parameters $B_1, \ldots, B_q, a_2, \ldots, a_q, b_1, \ldots, b_{q-1}$ can be estimated by regressing the frequencies at time $t+1$ on the frequencies at time t in a similar manner to the method of estimation that has been described for Lefkovitch's model. The problem of ill-conditioned estimates is still present, although there are fewer parameters. This problem was addressed by Caswell and Twombly (1989), who were concerned particularly with the problem of determining time changes in **M** matrices by estimating a series of these matrices based on data from successive sets of m generations from a longer series. They tried two standard methods for reducing variation in estimates, namely truncated singular value decomposition, and Tikhanov regularization (which is also known as ridge regression). They found truncated singular value decomposition the most useful of these two techniques. However, an unknown bias is introduced with its use and more work is required before it can be used with confidence.

Brown (1975) adopted a different approach to estimation in his study of the population dynamics of the freshwater pulmonate snail *Physa ampullacea*. He noted the need for non-negative coefficients and used linear programming to estimate the coefficients in row 1 of the transition matrix. This involved an arbitrary choice of coefficients in an objective function to be maximized, which seems to make it a generally unsatisfactory method. He used a form of least squares for estimating the other coefficients.

Example 6.1   Modelling Population Dynamics of Loggerhead Turtles

An interesting application of Usher's model has been concerned with the conservation of loggerhead sea turtles, *Caretta caretta*. Many populations are threatened with extinction and Crouse *et al.* (1987) considered the question of how numbers are related to survival and fecundity parameters in stages in order to determine the best parameters to change in order to increase population sizes. Seven life stages were considered: (1) eggs and hatchlings (under 1 year); (2) small juveniles (1–7 years); (3) large juveniles (8–15 years); (4) subadults (16–21 years); (5) novice breeders (22 years); (6) first-year remigrants (23 years); and (7) mature breeders (24–54 years).

The transition matrix that was assumed to apply for present populations is as follows:

$$
\begin{bmatrix}
0 & 0 & 0 & 0 & 127 & 4 & 80 \\
0.6749 & 0.7370 & 0 & 0 & 0 & 0 & 0 \\
0 & 0.0486 & 0.6610 & 0 & 0 & 0 & 0 \\
0 & 0 & 0.0147 & 0.6907 & 0 & 0 & 0 \\
0 & 0 & 0 & 0.0518 & 0 & 0 & 0 \\
0 & 0 & 0 & 0 & 0.8091 & 0 & 0 \\
0 & 0 & 0 & 0 & 0 & 0.8091 & 0.8089
\end{bmatrix}
$$

Reproduction is from the last three stages only, with fecundity rates being 127, 4 and 80 each year, respectively. It can also be seen, for example, that in stage 2 the probability of surviving and remaining in the stage from one year to the next is 0.7370, while the probability of surviving and moving to stage 3 is 0.0486. The dominant eigenvalue in this matrix is 0.9450, which indicates a long-term trend of a reduction in the total population size of about 5% a year.

It is straightforward with this model to determine the effect of changing one or more of the parameters in the transition matrix. Crouse *et al.*'s analysis shows that current conservation measures aimed at protecting eggs on nesting cannot be expected to halt population declines. Improving juvenile survival may be far more effective because a relatively small increase in survival rates in early stages would lead to an eigenvalue greater than 1.

Crouse *et al.* used a stage-based model for two reasons. First, much of the available demographic information was based on stages rather than ages. Second, an age-based model using a yearly time interval would have led to an unwieldy transition matrix with 54 rows and columns. To determine their transition matrix they assumed a geometric decline with age for numbers within a stage. The elements in the matrix relating to survival and promotion through stages could then be written in terms of yearly survival rates and the durations of stages. Crouse *et al.*'s paper should be consulted for information on how population characteristics were estimated.

6.5 FURTHER GENERALIZATIONS AND EXTENSIONS

Hiby and Mullen (1980) discussed the use of the Bernardelli–Leslie–Lewis model to estimate mortality rates in stages when age-specific fecundity or birth rates and (constant) stage durations are known. To do this they modified the model to

$$\mathbf{G}_{t+h} = \mathbf{B}_t \mathbf{L}_t^h \mathbf{U}_t$$

where $\mathbf{G}_{t+h}$ is a vector of stage-frequencies at time $t + h$, $\mathbf{B}_t$ is a matrix of zeros and ones that has the effect of converting age-frequencies to stage-frequencies, $\mathbf{L}_t$ is the usual Leslie matrix, and $\mathbf{U}_t$ is the age distribution at time t. Mortality rates are elements of the Leslie matrix and can be estimated if everything else is known. Given a sequence of $\mathbf{G}$ matrices it is possible to estimate a sequence of time-dependent survival rates. (See Hiby and Mullen's paper for more details.)

Woodward (1982) suggested using the Bernardelli–Leslie–Lewis model in an indirect way with stage-structured populations by setting up of an experimental cohort from birth (hatching) to death in order to determine survival rates and the proportions in each stage at each age. It then becomes possible to project population numbers forward in time in terms of ages, and see what stage distributions are implied by the age distributions. Woodward determines the distributions of stages using discrete approximations to normal distributions. The main limitation of his proposal is obviously the need for an experimental cohort with exactly the same survival rates and stage distributions as the population to be projected.

Longstaff (1984) has also considered an extension of the Bernardelli–Leslie–Lewis model. This has 30 immature age classes and 40 adult classes with the affect that the age of transition from immature to adult has a lognormal distribution. This reduces discontinuities in the population trajectory. The model was developed particularly for populations of the rice weevil *Sitophilus oryzae*.

Caswell (1986) has reviewed the use of matrix population models in the study of the complex life cycles of plants.

6.6   SAMPLING VARIATION AND OTHER SOURCES OF ERRORS

At best, the matrix models described above will only be approximations to reality. Even with very large populations it cannot be expected that stage-frequencies will exactly follow a course determined by a fixed transition matrix. Also, data stage-frequencies will usually only be estimates of population values so that even if a model is quite correct the data will not be fitted exactly. Hence, there is a need for model descriptions to incorporate allowances for random variation in population stage-frequencies and sampling errors.

Sykes (1969) discussed stochastic versions of the classical Bernardelli–Leslie–Lewis model. He suggested three plausible approaches. First, there can be additive errors put on the basic model so that equation (6.3) becomes

$$\mathbf{N_{t+1}} = \mathbf{MN}_t + \boldsymbol{\varepsilon}_t, \tag{6.9}$$

where $\boldsymbol{\varepsilon}_t$ is a random vector. Second, the elements of the transition matrix $\mathbf{M}$ can be thought of as probabilities rather than deterministic rates. Third, the elements of the transition matrix can be random variables, rather than constants.

Most attention in the biological literature has been towards the third of these approaches, which seems to be the most realistic (Boyce, 1977; Tuljapurkar and Orzak, 1980; Tuljapurkar, 1982; Slade and Levenson, 1982; Nordheim *et al.*, 1989). Perhaps the most interesting aspect of these studies has been the discovery that with random elements in the transition matrix the distribution of the total population size has a lognormal distribution after a population has been developing for a long time. This leads to the result that almost all simulated realizations of the development of a population may give population sizes that are lower than the expected size (Nordheim *et al.*, 1989). Hence, a population with a mean Leslie matrix implying a strong growth rate may in fact have a high probability of dying out.

The three ways for generalizing the Bernardelli–Leslie–Lewis model can also be used with matrix models for populations grouped in stages. Nordheim *et al.* (1989) modelled aphid populations the third way, using 15×15 stochastic transition matrices, assuming beta distributions for stage survival rates and gamma distributions for reproduction rates, with rates being independently distributed both within and across time periods. Manly (1989a) proposed a model for more general situations that takes into account both stochastic population variation and sampling errors.

The Manly (1989a) model says that for the population the relationship between the stage-frequencies at time t and at time $t+1$ is

$$
\begin{bmatrix} f_1(t+1) \\ f_2(t+1) \\ \cdot \\ \cdot \\ \cdot \\ f_q(t+1) \end{bmatrix} = \begin{bmatrix} m_{11} & m_{12} & \cdots & m_{1q} \\ m_{21} & m_{22} & \cdots & m_{2q} \\ \cdot \\ \cdot \\ \cdot \\ m_{q1} & m_{q2} & \cdots & m_{qq} \end{bmatrix} \begin{bmatrix} f_1(t) \\ f_2(t) \\ \cdot \\ \cdot \\ \cdot \\ f_q(t) \end{bmatrix} + \begin{bmatrix} \varepsilon_1(t+1) \\ \varepsilon_2(t+1) \\ \cdot \\ \cdot \\ \cdot \\ \varepsilon_q(t+1) \end{bmatrix},
$$

or

$$\mathbf{F}_{t+1} = \mathbf{M}\mathbf{F}_t + \boldsymbol{\varepsilon}_{t+1}. \tag{6.10}$$

This is Lefkovitch's model with a vector of 'errors' $\boldsymbol{\varepsilon}_{t+1}$, which takes into account random disturbances in the system. It is Usher's model with errors if the appropriate transition coefficients are set at zero. The mean vector for the errors is assumed to be a zero vector and $\varepsilon_i(t+1)$ and $\varepsilon_j(t+1)$ are assumed to be independent, $i \neq j$. Generally, it can be expected that the magnitude of the disturbances will increase with the population size and this can be accounted for by making the standard deviation of $\varepsilon_j(t+1)$ equal to the expected value of $f_j(t+1)$ multiplied by a constant α.

Obviously, this model may be too simple to account realistically for the stochastic variation in a real population. In particular, it may be that the errors show serial correlation over time, or correlation between stages at one time. However, it does have the merit of having only a single parameter to account for stochastic population variation. With most sets or real data it may be difficult to estimate more than one parameter for this purpose.

The parameter α can be estimated by equating the total of the residual sums of squares from the regressions of stage-frequencies at time $t+1$ to stage-frequencies at time t (Manly, 1989a). An alternative possibility is to simulate data with the model and find the value of α that gives the observed total residual sum of squares. This involves less approximation but is far more computer intensive.

Example 6.2   A Laboratory Population of *Lasioderma serricorne*

Consider again Lefkovitch's (1964a) experiment II on a laboratory population of the cigarette beetle *Lasioderma serricorne*. There were four replicate populations in this case, each started with 20 adults. Total stage-frequencies from the four replicates are given in Table 1.4 and the results for the individual replicates are provided in Table 6.1. The experiment was de-

Table 6.1 Replicate results from Lefkovitch's (1964a) second experiment (e = eggs, l = larvae, p = pupae, a = adults)

Sample time	e	l	p	a	e	l	p	a
		Replicate 1				Replicate 2		
0	0	0	0	20	0	0	0	20
3	0	432	0	0	0	356	0	0
6	1377	6	12	226	642	10	4	322
9	2	2230	0	8	0	1913	0	2
12	0	593	142	70	0	458	253	113
15	240	651	6	363	492	1067	6	186
18	54	979	111	107	1	631	85	3
21	19	679	15	78	442	442	19	215
24	430	589	25	259	15	1291	22	39
27	78	1444	0	130	11	504	124	85
30	20	695	14	76	672	962	17	185
33	392	290	25	289	26	934	32	70
36	100	1413	15	79	346	742	72	204
39	31	622	62	22	35	1181	59	121
42	592	428	28	193	48	819	16	107
45	30	1709	18	42	291	835	38	256
48	24	723	46	36	69	1152	60	104
		Replicate 3				Replicate 4		
0	0	0	0	20	0	0	0	20
3	0	250	0	0	0	317	0	0
6	1132	11	4	218	1520	26	3	242
9	1	1519	2	2	0	1883	0	0
12	1	247	199	315	116	288	67	285
15	302	1655	1	117	162	1376	0	189
18	0	699	112	2	0	529	243	23
21	537	543	9	162	1025	824	11	249
24	5	1121	26	19	0	1116	9	11
27	311	431	84	162	601	522	80	282
30	29	806	11	10	126	1453	12	175
33	108	186	39	84	8	424	158	73
36	45	616	6	31	686	1112	19	212
39	85	244	96	129	5	890	61	25
42	70	1227	18	27	496	694	60	267
45	33	498	45	84	35	2216	28	75
48	223	976	21	94	42	663	91	41

signed to make the replicate populations emulate natural populations with ample food but limited space. There are four stages (eggs, larvae, pupae and adults).

Inspection of the data indicates that the time between observations exceeds at least some of the stage durations. For example, it can be seen that the adults present in the first sample produced larvae by the time of the second sample, but no eggs remained. The larvae were then able to produce individuals in all stages by the time of the third sample. It seems to be the case that the egg, pupae and adult stages take less than the 3-week period between samples, but the larvae stage takes more than 3 weeks. On this basis, the form of the **M** matrix of equation (6.6) is as follows:

$$
\begin{array}{c}
 \\
\text{Eggs (e)} \\
\text{Larvae (l)} \\
\text{Pupae (p)} \\
\text{Adults (a)}
\end{array}
\begin{array}{cccc}
e & l & p & a \\
\left[\begin{array}{cccc}
0 & m_{12} & m_{13} & 0 \\
m_{21} & m_{22} & m_{23} & m_{24} \\
0 & m_{32} & 0 & 0 \\
0 & m_{42} & m_{43} & 0
\end{array}\right] &&&
\end{array}.
$$

In all four replicate populations there was an initial fast growth in population numbers, due presumably to the relative abundance of food, followed by oscillations around a general level of about 1500 *L. serricorne*. The initial fast-growth phase was over by the time of the third sample and it seems appropriate to omit the first two samples from the data for estimation purposes. Based on the remaining 15 samples, the multiple regression method provides estimated **M** matrices for the four replicates that are shown in Table 6.2. The means of the transition coefficients $\pm$ their estimated standard errors are as follows:

$$
\left[\begin{array}{cccc}
0 & 0.04 \pm 0.01 & 2.56 \pm 0.47 & 0 \\
0.67 \pm 0.29 & 0.17 \pm 0.04 & 3.28 \pm 0.51 & 3.21 \pm 0.58 \\
0 & 0.07 \pm 0.01 & 0 & 0 \\
0 & 0.07 \pm 0.01 & 1.04 \pm 0.30 & 0
\end{array}\right].
$$

Because the standard errors are based on only four replicates they are not very reliable. Nevertheless, it does seem that the transition parameters are estimated with fair accuracy.

The residual sums of squares for the four replicates are 1024430, 1069990, 790529 and 2193700, respectively, with a mean of 1269662. Simulations using the mean estimated transition matrix, starting with 1168 in stage 1, 13 in stage 2, 6 in stage 3 and 252 in stage 4 (the average sample-3 composition of the four replicates), with a range of values for α, show that if $\alpha = 0.45$ then the mean residual sum of squares is about

Table 6.2   Regression estimates of transition matrices for the four replicate populations of Lefkovitch's (1964a) experiment II

Replicate 1					Replicate 2			
0	0.04	1.95	0		0	0.03	2.60	0
1.41	0.26	2.55	1.92		0*	0.15	2.27	4.75
0	0.04	0	0		0	0.08	0	0
0	0.05	1.84	0		0	0.08	0.83	0

Replicate 3					Replicate 4			
0	0.06	1.84	0		0	0.03	3.88	0
0.73	0.17	4.09	2.98		0.56	0.08	4.23	3.19
0	0.08	0	0		0	0.07	0	0
0	0.10	0.41	0		0	0.07	1.09	0

*The value indicated for replicate 2 was estimated to be negative. It was therefore set at zero and the other parameters re-estimated.

equal to the mean value for the replicates. This can therefore be taken as the estimate of α appropriate for the data.

When 100 populations were simulated using $\alpha = 0.45$ and other conditions as just mentioned, and the data obtained were analysed in the same way as the real data, the mean transition matrix obtained was as follows:

$$\begin{bmatrix} 0.00 & 0.04 & 2.46 & 0.00 \\ 0.83 & 0.16 & 3.30 & 2.25 \\ 0.00 & 0.06 & 0.00 & 0.00 \\ 0.00 & 0.06 & 0.97 & 0.00 \end{bmatrix}.$$

The transition matrix used for the simulation has been more or less recovered. The standard deviations of the 100 estimates obtained were as shown below:

$$\begin{bmatrix} 0.00 & 0.02 & 0.78 & 0.00 \\ 0.64 & 0.10 & 2.61 & 1.69 \\ 0.00 & 0.02 & 0.00 & 0.00 \\ 0.00 & 0.02 & 0.41 & 0.00 \end{bmatrix}.$$

Standard errors for the replicate means can be estimated by dividing the standard deviations by two (the square root of the sample size). The results obtained from doing so are reasonably close to the standard errors given above that were calculated directly from the replicates.

The dominant eigenvalue of the matrix of mean estimated transition coefficients is 0.9961, with the corresponding eigenvector being [0.2124,

1.0000, 0.0666, 0.1446]. This suggests that the populations set up by Lefkovitch would have declined very slowly in the long term if the experiment had been continued. However, as mentioned before, studies of stochastic Leslie matrices indicate that the behaviour of individual populations can be quite different from what is found when stochastic variation is ignored. A better indication of likely behaviour would be given by simulating many populations for long periods of time, but this matter will not be pursued here.

6.7 DISCUSSION

When Lefkovitch introduced his model for populations developing through stages he envisaged a situation where samples are taken at time intervals that are fairly large in comparison with the mean durations of stages. This is the reason why none of the elements in the transition matrix is necessarily zero with his model. This gives great flexibility. However, the elements in the matrix do not have clear interpretations. Usher's model is much more satisfactory in this respect because it allows the elements to be interpreted in terms of birth rates and survival probabilities. This suggests that if data are going to be collected for analysis with a matrix model then every effort should be made to ensure that the time between observations is shorter than the minimum duration of any stage.

There is a problem with modelling populations using transition matrices that has been mentioned above but needs some further discussion. This concerns the assumption that the same transition matrix applies, irrespective of the age distribution of individuals within stages. There are two situations for which this will hold. Either the distributions of stage durations can be exponential, so that the probability of moving to the next stage in the time between samples is the same for individuals of all ages, or there has to be a stable distribution of ages within stages. The first of these situations is unlikely to be true as it can be expected that a definite maturation process takes place in each stage, and the probability of an individual leaving a stage depends on how long the individual has been in that stage. If a population is in a fairly stable state as far as the age distribution within stages is concerned then a transition matrix model may give a reasonable fit to data. The possibility of a bad fit to data because of a changing age distribution should always be kept in mind.

6.8 OTHER MODELS FOR POPULATIONS WITH CONTINUOUS RECRUITMENT

Although completely stable stage-structures seem to occur rarely in natural populations, several methods for estimating population parameters have been proposed based on this assumption for populations with continuous recruitment.

Hughes (1962, 1972) noted that a population can be expected to grow exponentially with physiological time, with a constant ratio for the numbers in different stages, once the generations overlap. He pointed out that with aphids the potential rate of increase per instar period can be estimated as the natural logarithm of the numbers in successive instars when the instar periods are of the same length. The mortality rate and the recruitment rate can also be estimated. Hughes also provided evidence of stable stage distributions for wild populations of five aphid species. Hughes's method was used for the International Biological Programme's study of the aphid *Myzus persicae* (Mackauer and Way, 1976). However, Carter *et al.* (1978) have suggested that in fact the model has limited use with wild aphid populations as these populations seldom achieve the required stable stage proportions.

Many authors have been particularly concerned with estimation for zooplankton populations using the ratio of eggs to adults. Seitz (1979) proposed a two-stage (egg–adult) model, with the same death rate in each stage, and a three-stage (egg–young–adult) model with equal death rates for the first and last stage. These are generalizations of methods proposed earlier by Edmonson (1960, 1968), Caswell (1972) and Paloheimo (1974). From a simulation study, Seitz concluded that it is best to use his two-stage model when the mortality rate is the same in both stages. When mortality rates change from stage to stage, Paloheimo's model gives the best results. However, Seitz warns that the results from all the methods must be treated with caution when there are strong deviations from the steady state in the population being analysed.

More recently, Gabriel *et al.* (1987) have compared the methods of Edmonson (1960), Caswell (1972), Paloheimo (1974), Seitz (1979) and Taylor and Slatkin (1981) using data from 37 laboratory populations of the copepod crustacean *Daphnia pulicaria* and computer simulations. For sampling intervals equal to the duration of the egg stage (2.5 days) Paloheimo's formula for birth rates was most accurate but for longer intervals (7.5–10 days) Taylor and Slatkin's equation (22) was better. See also Taylor (1988).

Dorazio (1986) addressed the problem of estimation with populations with changing birth rates. He proposed sampling frequently enough so that corrections for the exponential growth of a population are not necessary. The change in the population size from one sample time to the next can then be used to estimate the difference between the birth rate and the death rate per unit time. The birth rate applying over the interval between two samples can be estimated directly from the equation

$$B = (EH)/(Ns),$$

where E is the number of eggs at the start of the interval, H is an estimate of the fraction of these eggs that hatch, N is the population time at the start of the interval and s is the length of the interval. The value of H is estimated either by incubating a random sample of eggs or by assigning eggs to age classes and calculating the proportion that should hatch by the end of the interval.

Paloheimo's formula has found favour in recent discussions (Hart, 1987). This estimates the instantaneous birth rate in a population as

$$b = \log_e(E/N + 1)/D, \tag{6.11}$$

Table 6.3   Artificial data generated using the model of equations (6.10)

Sample time	Frequency in stage			
	1	2	3	4
1	924	0	0	0
2	532	379	0	0
3	251	412	141	0
4	133	276	252	51
5	107	188	248	137
6	201	122	188	211
7	234	125	177	231
8	329	180	142	236
9	433	231	155	287
10	407	254	135	265
11	523	333	167	253
12	498	354	203	255
13	490	332	227	255
14	555	415	253	349
15	553	409	303	349
16	646	387	313	381
17	709	519	308	432
18	721	526	328	556
19	921	581	356	538
20	947	654	311	570
21	1056	653	393	587
22	1119	722	481	622
23	1067	748	560	716
24	1270	813	543	784
25	1344	922	634	818
26	1411	991	584	838
27	1621	1052	803	838
28	1521	1219	845	923
29	1779	1234	892	1121
30	1906	1277	896	1261

where E is the observed egg density (the number of egg-bearing females $\times$ the mean clutch size + detached eggs or embryos), N is the density of post embryos and D is the (known) embryonic duration at the prevailing temperature. The formula is based on the assumption of constant birth and death rates, and a fixed time for egg development. It was first mentioned by Edmonson (1968), but justified more fully by Paloheimo (1974).

Assuming exponential population growth, so that the relationship between the population size at time 0 and time t is $N_t = N_0 \exp(rt)$, the instantaneous population growth rate r can be estimated from the equation

$$r = \log_e(N_t/N_0)/t. \tag{6.12}$$

The instantaneous death rate can then be estimated from the equation

$$d = b - r. \tag{6.13}$$

Equations (6.12) to (6.14) constitute Paloheimo's method for estimating population parameters for a stable population.

Keen and Nassar (1981) have addressed the question of the determination of the accuracy of the estimates from these equations. They show that r, b and d should have approximately normal distributions if sample counts for eggs and adults have Poisson or negative binomial distributions. They propose the determination of the accuracy of estimates from replicated field samples.

EXERCISE

Table 6.3 shows some artificial data that were obtained using the model of equation (6.10) with the transition matrix

$$\begin{bmatrix} 0.5 & 0.0 & 0.0 & 1.0 \\ 0.4 & 0.5 & 0.0 & 0.0 \\ 0.0 & 0.4 & 0.5 & 0.0 \\ 0.0 & 0.0 & 0.4 & 0.8 \end{bmatrix},$$

and an α value of 0.05. These are sample data, with Poisson sampling errors superimposed on expected frequencies for the population. Use regression methods to estimate the transition matrix and the α value. See how close the estimates are to the values used to generate the data.

7 Key factor analysis

7.1 POPULATIONS OBSERVED OVER A SERIES OF GENERATIONS

For populations with distinct generations it is often possible to count or estimate the total number of individuals entering different development stages for a series of generations using the methods described in previous chapters. The variation in the survival rates in different stages from generation to generation can then be studied in order to gain understanding of which sources of variation are particularly important for population dynamics. It is this type of investigation that is the subject of the present chapter.

An example set of data is shown in Table 7.1. Here the population is the winter moth *Operophtera brumata* in Wytham Wood, near Oxford, England. Estimates of the numbers of moths at seven distinct points in the life cycle are available for each of the years 1950–66, with each year providing the life table for one generation. The first count in each year is the maximum number of eggs that could be produced by the adults in the previous year. The last count in each year is the individuals that survive to become adults. From the counts for each generation it is possible to calculate the following survival rates through life stages: overwinter survival, survival of parasitism I, survival of parasitism II, survival of parasitism III, survival of pupal predation and survival of pupal parasitism.

Table 7.2 gives another example. In this case the population is the tawny owl *Strix aluco*, also in Wytham Wood. Numbers entering six developmental stages are given, starting with the maximum possible number of eggs from the adult pairs, and ending with the young surviving until the next year.

In considering the analysis of this type of data it is important to realize that variation in survival rates is in some respects more important than mean survival rates. This was first emphasized by Morris (1957, p. 57) who expressed it cogently in the following words:

> By variation in mortality I refer to the significant changes that occur from generation to generation . . . not the minor variations associated with sampling errors. The first point to be established is that it is this variation in mortality rather than the absolute level of mortality that is the important thing in population dynamics. This point is obvious, of course, for we would not have population dynamics unless mortality . . . was variable. For example, there is a mortality of 10% affecting the eggs of the spruce budworm which may be attributed to such things as infertility and

hatching failure. Its variation with place or time during our studies has been slight. It has therefore had no appreciable influence on population changes. We could, in fact, leave it out of life tables and say that the mean fecundity is 180 instead of 200. The same argument would apply if the degree of mortality were 90% instead of 10%; as long as it does not vary it does not contribute to the changes in population, but only to the potential rate of increase or decrease of the species. For this reason a single life table applying to one generation of an insect . . . reveals little about the population dynamics of the insect. Nor can it be used to indicate the relative importance of the different mortality factors because some of the highest mortalities may be relatively constant while some of the low ones may vary greatly.

Table 7.1   Life tables for the winter moth *Operophtera brumata* in Wytham Wood, Berkshire, as determined from Table F and Figure 7.4 of Varley *et al.* (1973). All frequencies are per square metre

	Number of survivors to stage						
Year	*1*	*2*	*3*	*4*	*5*	*6*	*7*
1950	4365.0	112.2	87.1	79.4	70.8	14.5	7.41
1951	417.0	117.5	114.8	109.6	102.3	17.4	13.8
1952	758.6	55.0	54.6	47.5	42.4	7.50	7.03
1953	389.0	18.2	17.8	17.0	15.5	7.08	4.90
1954	275.4	158.5	157.0	146.6	127.6	23.8	20.2
1955	1122.0	77.6	77.1	71.9	65.6	14.7	11.9
1956	645.7	95.5	89.1	87.0	83.2	28.2	14.8
1957	831.8	275.4	263.0	257.0	229.1	37.2	23.4
1958	1288.0	190.5	162.2	154.9	141.3	21.4	14.8
1959	812.8	57.5	45.7	41.7	39.8	8.71	6.17
1960	346.7	21.4	20.4	17.8	16.6	3.16	1.12
1961	61.7	7.59	7.59	6.61	6.03	3.63	3.02
1962	166.0	13.5	12.9	11.2	10.5	6.03	5.25
1963	288.4	40.7	40.7	37.2	36.3	14.5	11.0
1964	602.6	131.8	130.3	127.4	124.5	22.6	16.4
1965	891.3	269.2	251.2	245.5	239.9	32.4	24.0
1966	1349.0	51.3	46.8	46.2	44.2	6.10	2.85
1967	154.9	9.77	9.55	8.91	8.91	2.82	2.82
1968	154.9	10.0	10.0	9.77	9.12	3.02	3.02

Stages are: (1) Egg output from the adults in the previous year, which is taken by Varley *et al.* to be 56.25 eggs per adult; (2) fully grown larvae; (3) after parasitism by *Cyzensis albicans*; (4) after parasitism by non-specific Diptera and Hymenoptera; (5) after parasitism by *Plistophora operophterae*; (6) after pupal predation; (7) adults, after parasitism by *Cratichneumon culex*.

Table 7.2   Life tables for the tawny owl *Strix aluco* in Wytham Wood, as given by Southern (1970). No pairs bred in 1958

Year	Numbers in stages					
	1	*2*	*3*	*4*	*5*	*6*
1949	60	54	50	34	26	7
1950	75	66	57	31	25	9
1951	63	36	25	6	6	6
1952	72	51	43	16	21	9
1953	75	48	32	20	20	13
1954	84	60	50	15	17	17
1955	90	12	8	4	4	4
1956	99	66	48	23	24	12
1957	102	66	66	26	20	10
1958	93	0	0	0	0	0
1959	110	81	65	25	28	12

Stages are: (1) maximum number of eggs possible for the adult pairs present; (2) maximum number of eggs possible for breeding pairs; (3) eggs produced; (4) eggs hatched; (5) young fledged; (6) young remaining next spring. In some years, sampling errors have resulted in stage 5 numbers being higher than stage 4 numbers.

This argument, which can be made equally well in terms of survival rates rather than mortality rates, shows that there is value in comparing levels of variation in survival rates in different stages, and it is also important to see how the variation in survival rates in one stage affects numbers in later stages. **Key factor analysis** is the term that has come to be used for this type of study, where the key factor is the survival in the stage of the life cycle that contributes most to variation in overall population numbers. A major goal is to identify this stage. Another important consideration may be the identification of the stages for which the survival rates are so variable as to make it necessary to study them further in order to develop a realistic model for the population. The stage-specific survival rates that vary little can be treated as being constant, at least in the initial stages of model building. The further study of stages with variable survival rates involves attempting to relate this variation to environmental variables such as climate, numbers of predators, number of parasites or the availability of food. In this way, it can be hoped that a clearer understanding of the major causes of population changes will emerge.

This chapter begins with a review of the methods that have been proposed for key factor analysis. In order to simplify this review, the effects of sampling errors in data and questions concerning sampling distributions of estimates are left until later in the chapter.

Some important ideas must be mentioned at this point in order to put key factor analysis into the correct perspective. The first of these is that there may be very little relationship, if any, between the fecundity of species and the size of populations. In many cases, almost all individuals die before reaching maturity. Hence, the survival rate in a particular life stage may be far more important to population dynamics than the fecundity rate.

A second point to note is that some populations are closed in the sense that the number entering stage 1 in one generation can be calculated directly from the number in the last stage in the previous generation. Typically, this is done by assuming some potential fecundity rate. The winter moth data in Table 7.1 give a situation like this. Varley *et al.* (1973) used a fecundity rate of 56.25 eggs per adult. Hence, for example, the figure of 417.0 eggs in 1951 is (apart from a rounding error) simply the adults in the previous year 7.41 multiplied by 56.25. In effect, this means that the first survival rate calculated in each generation (fully grown larvae/maximum possible eggs), which is called the overwinter loss, includes components for lowered fecundity, the loss of eggs and early larval mortality.

On the other hand, cases do arise where the relationship between the numbers in successive generations is more obscure. The individuals entering stage 1 in each generation may be produced largely or partly by immigrant adults, or there may be contributions from several previous generations. In that case, it may be better to regard the number of individuals entering stage 1 as a source of variation that is independent from generation to generation. This is the situation with the tawny owl data in Table 7.2. Here, the maximum number of eggs possible was calculated from the number of pairs of owls present in the study area, where these owls consisted of adults surviving from the previous year, fledglings from the previous year that survived and remained in the area, and adults that immigrated from outside the area.

7.2   DENSITY-DEPENDENT SURVIVAL

Varley and Gradwell (1970) note that four terms, **density-dependent mortality**, **inverse density-dependent mortality**, **delayed density-dependent mortality** and **density-independent mortality**, arose from early theoretical ideas about population dynamics. They then gave rise to so many arguments about the classification of different effects that many people avoided their use altogether. However, with key factor analysis these terms are more or less essential so their definitions will be given here.

A density-dependent mortality rate, sometimes described as a negative feedback, is defined as one that increases as the population density increases. Thus, a large population has a high death rate and a small population a low death rate. This type of mortality obviously tends to stabilize population numbers, and may be quite common in nature. A decrease in the fecundity rate with increasing population size has the same effect.

An inverse density-dependent mortality, sometimes described as a positive feedback, is one for which the death rate decreases with increasing population density. Clearly, this will have the effect of destabilizing a population. An increasing fecundity rate with increasing population size has the same effect. What may occur in some cases is that a mortality rate is inversely density-dependent for small population sizes but becomes density-dependent for high population sizes.

A delayed density-dependent mortality is one for which the effects of population density do not occur immediately. The example usually quoted is for the situation where the mortality is associated with a parasite or predator (Nicholson, 1933; Nicholson and Bailey, 1935). A large population size in one generation for the organism being studied will then encourage the growth of the parasite or predator population in the next generation. Consequently, the mortality rate will tend to be high in the generation following a generation with a large population size.

A density-independent mortality is one that is not related in any way to population density. Often, this will be mortality associated with climatic effects, food supplies and many 'random' effects. In practice, some part of the variation in all mortality rates can be expected to be density-independent.

7.3   THE VARLEY AND GRADWELL (1960) GRAPHICAL METHOD FOR KEY FACTOR ANALYSIS

The idea of a key factor analysis arose from the classical papers of Morris (1957, 1959) and Varley and Gradwell (1960). The graphical method suggested by Varley and Gradwell seems to have been particularly influential.

The simplicity of the ideas put forward by Varley and Gradwell is undoubtedly responsible for their popularity. Thus, consider one generation of a population, for which the numbers entering developmental stages $1-q$ are $N_1, N_2, \ldots, N_q$. The stage-specific survival rates are then $w_1 = N_2/N_1$, $w_2 = N_3/N_2, \ldots, w_{q-1} = N_q/N_{q-1}$, and the total survival to the adult stage is

$$N_q/N_1 = w_1 w_2 \ldots w_{q-1}.$$

Taking logarithms (base 10 being usually used) then gives

$$\log(N_q/N_1) = \log(w_1) + \log(w_2) + \ldots + \log(w_{q-1}),$$

or

$$K = k_1 + k_2 + \ldots + k_{q-1},$$

where $K = -\log(N_q/N_1)$ and $k_j = -\log(w_j)$. The changes of sign are simply made to avoid negative values. Varley and Gradwell suggested plotting K and the k values against the same time axis for a series of generations. The 'key factor' is then the stage for which variation in the k value most closely follows variation in K, as determined by a visual inspection.

Example 7.1   Graphical Key Factor Analysis of the Winter Moth Data

The K and k values for the winter moth data of Table 7.1 are shown in Table 7.3, and plotted on Figure 7.1. The key factor seems to be k_1, the overwinter loss, as it is the plot of k_1 that is most similar to the plot of K. The k_5 value, reflecting loss of pupae in the soil due to predation, is also quite variable.

Table 7.3   Calculated K and k values for the winter moth data of Table 7.1 using base 10 logarithms

		k values					
Year	K	1	2	3	4	5	6
1950	2.77	1.59	0.11	0.04	0.05	0.69	0.29
1951	1.48	0.55	0.01	0.02	0.03	0.77	0.10
1952	2.03	1.14	0.00	0.06	0.05	0.75	0.03
1953	1.90	1.33	0.00	0.02	0.04	0.34	0.16
1954	1.13	0.24	0.00	0.03	0.06	0.73	0.07
1955	1.97	1.16	0.00	0.03	0.04	0.65	0.09
1956	1.64	0.83	0.03	0.01	0.02	0.47	0.28
1957	1.55	0.48	0.02	0.01	0.05	0.79	0.20
1958	1.94	0.83	0.07	0.02	0.04	0.82	0.16
1959	2.12	1.15	0.10	0.04	0.02	0.66	0.15
1960	2.49	1.21	0.02	0.06	0.03	0.72	0.45
1961	1.31	0.91	0.00	0.06	0.04	0.22	0.08
1962	1.50	1.09	0.02	0.06	0.03	0.24	0.06
1963	1.42	0.85	0.00	0.04	0.01	0.40	0.12
1964	1.57	0.66	0.00	0.00	0.01	0.74	0.14
1965	1.57	0.52	0.03	0.00	0.01	0.87	0.13
1966	2.68	1.42	0.04	0.00	0.02	0.86	0.33
1967	1.74	1.20	0.00	0.03	0.00	0.50	0.00
1968	1.71	1.19	0.00	0.01	0.03	0.48	0.00
Mean	1.82	0.97	0.03	0.03	0.03	0.62	0.15
SD	0.45	0.36	0.03	0.02	0.02	0.21	0.12

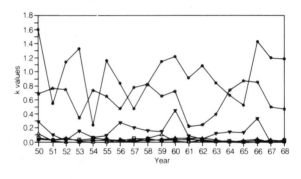

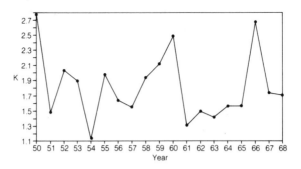

Figure 7.1   Key factor analysis for the winter moth data of Table 7.1. ●, k1; △, k2; □, k3; ▲, k4; ⋆, k5; ▽, k6.

Example 7.2   Graphical Key Factor Analysis of the Tawny Owl Data

The K and k values for the tawny owl data of Table 7.2 are shown in Table 7.4, and plotted on Figure 7.2. The plot of k_1 (loss because of birds failing to breed) follows the plot of K quite closely so that this seems to be the key factor here. The plot of k_5 (the death and emigration of young owlets after they leave the nest) is to some extent a mirror image of the plot of k_1, with high values of k_1 occurring with low values of k_5, and vice versa. This is an indication of density-dependent mortality in stage 5.

7.4   EXTENSIONS TO THE VARLEY AND GRADWELL APPROACH

The graphical model of Varley and Gradwell does often identify the key factor but unclear cases do arise. This has led several authors to propose more objective methods for evaluating the importance of different k values. In some cases, the correlations between the k values and K have been calculated with the highest value taken to indicate the key factor. In other cases, regressions of K on k values have been considered. However, Podoler

Table 7.4 Calculated *K* and *k* values for the tawny owl data of Table 7.2 using base 10 logarithms

| Time | K | k values | | | | |
		1	2	3	4	5
1949	0.93	0.05	0.03	0.17	0.12	0.57
1950	0.92	0.06	0.06	0.26	0.09	0.44
1951	1.02	0.24	0.16	0.62	0.00	0.00
1952	0.90	0.15	0.07	0.43	−0.12	0.37
1953	0.76	0.19	0.18	0.20	0.00	0.19
1954	0.69	0.15	0.08	0.52	−0.05	0.00
1955	1.35	0.88	0.18	0.30	0.00	0.00
1956	0.92	0.18	0.14	0.32	−0.02	0.30
1957	1.01	0.19	0.00	0.40	0.11	0.30
1959	0.96	0.13	0.10	0.41	−0.05	0.37
Mean	0.947	0.221	0.099	0.365	0.008	0.254
SD	0.175	0.238	0.061	0.141	0.078	0.201

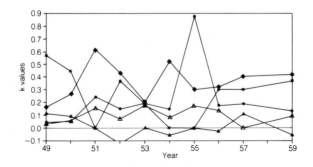

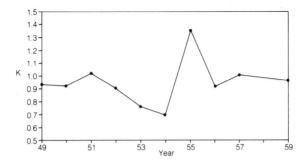

Figure 7.2 Key factor analysis for the tawny owl data of Table 7.2. ●, k1; △, k2; ◊, k3; ▲, k4; ★, k5.

and Rogers (1975) have pointed out that the best approach along these lines is to look at regressions of k values against K and use the regression slopes to compare the importance of different k values in determining K. Their method will therefore be discussed in more detail than the other ones.

The regression slope for k_j regressed on K is given by the usual equation

$$b_j = \sum_{i=1}^{G} (k_{ij} - \bar{k}_j)(K_i - \bar{K}) / \sum_{i=1}^{G} (K_i - \bar{K})^2, \qquad (7.1)$$

where k_{ij} is the value of k for the survival in stage j in generation i, $\bar{k}_j$ is the mean k value for survival in stage j, K_i is the K value in generation i, $\bar{K}$ is the mean of K for all generations, and there are G generations of data. It follows from this equation that

$$\sum_{j=1}^{q-1} b_j = 1.$$

Thus, the b_j values have the immediate interpretation of indicating the fraction of the total variation of K, as measured by $\Sigma(K_i - \bar{K})^2$, that is accounted for by k_j.

Although the idea of using the regression coefficients of k values on K to distinguish key factors is usually attributed to Podoler and Rogers, the same type of analysis was suggested much earlier by Henderson and Hayman (1960) in the context of studying the influence of fleece characters on wool production. The relevance of Henderson and Hayman's work was recognized by Young and Wrensch (1981) who used it for what was essentially a key factor analysis on experimental populations of spider mites. Young and Wrensch also noted the relevance of other statistical analyses used in wool assessment that are reviewed by Turner and Young (1969, ch. 14). Furthermore, what amounts in essence to the Podoler and Rogers analysis was also proposed by Smith (1973). He suggested measuring the relative importance of k_j by the covariance of k_j with K. This covariance is, of course, just b_j multiplied by a constant.

Smith also suggested a further analysis after the key factor has been found with the aim being to ascertain the relative importance of the remaining k values. Thus, suppose that k_p is the key factor. The residual 'killing power' in stages other than p is then $K_i' = K_i - k_{ip}$ in the ith generation, and the contribution of stage j to this residual is indicated by the covariance of K' and k_j. This allows the second most important k value to be determined. Continuing in this way, removing the k values one at a time, allows all the k values to be ranked in terms of importance. The calculations can be done either in terms of the covariances, as proposed by Smith, or by considering regression coefficients of k values on residual killing powers. In the latter case, the importance of k_j in determining K' is simply taken as the regression coefficient for k_j regressed on K', these coefficients adding to one.

The combination of using the regression coefficients to indicate the importance of different k values and extracting them one by one will be referred to from now on as the Podoler–Rogers–Smith method of key factor analysis.

Example 7.3   Regression Coefficients for the Winter Moth Data of Table 7.1

For the winter moth data of Table 7.1 the regression coefficients b_j are $b_1 = 0.62$, $b_2 = 0.04$, $b_3 = 0.00$, $b_4 = 0.00$, $b_5 = 0.16$ and $b_6 = 0.17$. The stage 1 survival is confirmed as the clear key factor. Removing k_1 values from K and calculating regression coefficients for the remaining k values on their total, along the lines suggested by Smith, indicates that the second most important factor is k_5. Removing this from the total and calculating regression coefficients indicate that the third most important factor is k_6, etc. The results of these calculations are summarized below:

| Step | Key factor | Stage | | | | | |
		1	2	3	4	5	6
1	1	0.62	0.04	0.00	0.00	0.16	0.17
2	5	0.00	0.06	−0.02	0.00	0.65	0.30
3	6	0.00	0.15	0.02	0.02	0.00	0.81
4	2	0.00	0.57	0.23	0.20	0.00	0.00
5	3	0.00	0.00	0.57	0.43	0.00	0.00

Example 7.4   Regression Coefficients for the Tawny Owl Data of Table 7.2

For the tawny owl data of Table 7.2 the results of the regression calculations are as shown below:

| Step | Key factor | Stage | | | | |
		1	2	3	4	5
1	1	1.08	0.08	−0.04	0.08	−0.20
2	5	0.00	−0.27	−0.05	0.23	1.08
3	3	0.00	0.13	0.95	−0.07	0.00
4	4	0.00	0.29	0.00	0.71	0.00

The coefficient $b_1 = 1.08$ in row 1 indicates that the values of k_1 (loss through failure to breed) are more variable than K. The negative coefficient $b_5 = -0.20$ shows that the survival in stage 5 (overwinter loss) has acted to reduce this variation. Once k_1 is removed from K, k_5 itself becomes the most important factor (row 2 of the table). The third most important factor is then found to be k_3 (loss due to failure of eggs to hatch). Lastly, we see that k_4 (loss due to young dying in the nest) seems to be more important than k_2 (loss due to the number of eggs laid by breeding pairs).

7.5   DETECTING DENSITY-DEPENDENT k VALUES

Following the establishment of key factors, the usual practice has been to graph k values against the logarithms of the population densities on which they act in order to see if there is any evidence for density-dependent survival. A positive relationship between k_j and log(density) is what indicates this. Plots for the winter moth data of Table 7.1 are shown on Figure 7.3, and for the tawny owl data of Table 7.2 on Figure 7.4. In the former case, there does seem to be density-dependence for k_5 (pupal predation), with a fitted regression line of

$$k_5 = 0.10 + 0.31 \log(\text{density}).$$

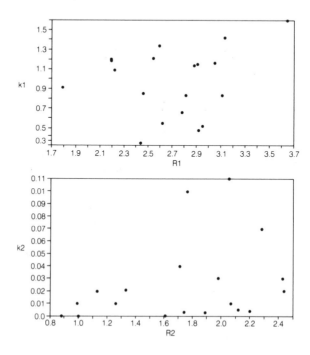

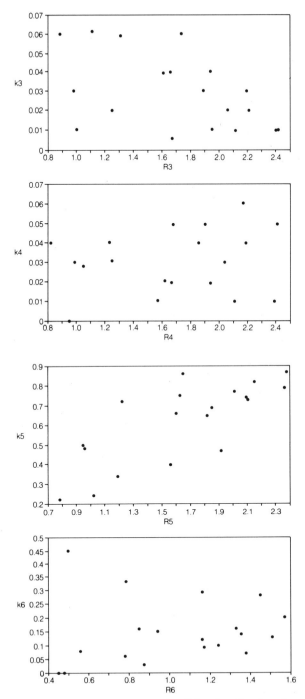

Figure 7.3 Plots of k values against $R = \log(N)$ values for the winter moth data of Table 7.1 to see if there is any evidence for density-dependent survival.

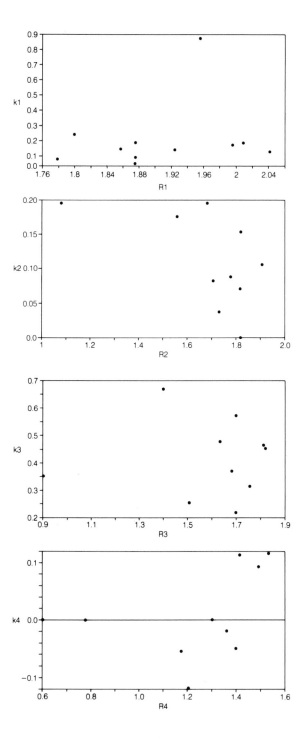

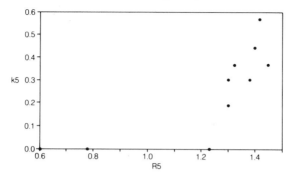

Figure 7.4 Plots of *k* values against *R* values for the tawny owl data of Table 7.2 to see if there is any evidence for density-dependent survival.

In the case of the tawny owl, it is also k_5 that appears to be density-dependent, with a fitted regression line

$$k_5 = -0.41 + 0.54 \log(\text{density}).$$

Unfortunately, testing regression lines like this for statistical 'significance' is not straightforward because sampling errors in *k* values may well induce apparent regression relationships (Kuno, 1971; Ito, 1972; Slade, 1977), and even if there are no sampling errors the standard regression model may not be appropriate (Royama, 1977). This matter is considered further in Section 7.8. Here, we can merely note that ordinary regression methods are not appropriate except under special circumstances. The methods discussed by Reddingius and den Boer (1989) and den Boer and Reddingius (1989) are also relevant here, although they do not specifically take into account stage structure.

Of course, the relationship between *k* values and population densities may be something more complicated than a linear regression. Indeed, the fact that negative *k* values are impossible suggests this immediately. Usually, the data do not really warrant fitting a more complicated model. However, one possible alternative to the linear model, that was suggested by Hassell (1975), is $\log(k) = b \log(1 + aN)$, where *a* and *b* are constants, and *N* is the density on which *k* acts.

As noted in Section 7.2, it may be that predation or parasitism on a population will cause delayed density-dependent mortality, with a high density in one generation leading to high predator numbers in the next generation. This is liable to show up in an anticlockwise circular or spiral graph when *k* values for successive generations are joined, as indicated in Figure 7.5 (Varley *et al.*, 1973, pp. 64, 124). A pattern of this type is suggested for k_1 and k_5 with the winter moth data (Figure 7.6) so it seems that this may be an example where this occurs. The question of how one

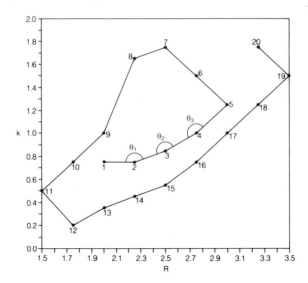

Figure 7.5   A spiral plot of *k* values against *R* = log (*N*) values indicating delayed density-dependent mortality. Generation numbers are shown.

measures 'spiralness', and determines whether an observed amount is significant, is considered in Section 7.9.

7.6   THE MANLY (1977) MODEL FOR KEY FACTOR ANALYSIS

A limitation with the methods discussed so far in this chapter is that they do not directly make use of the known order in which mortality operates through stages. However, this order is crucial if density-dependent mortality occurs. For example, it can happen that a highly density-dependent mortality in a stage in the life cycle removes almost all of the variation introduced by variable mortality in earlier stages. The density-dependent *k* values may then show little relationship, if any, to the *K* values. Nevertheless, the stage in question is extremely important for population dynamics.

One way to uncover a 'hidden' key factor like this involves estimating the parameters in a population model that allows for density-dependent mortality. From these parameters it is then possible to estimate the contribution of each life stage to variation in the numbers entering the final stage. A key factor can then be defined as a life stage that substantially either increases or reduces this variation.

An approach from this point of view was suggested by Manly (1977b, 1979). The notation used here will be changed from that originally used in order to fit in better with the equations already used in this chapter. As before, let $k_j = -\log(N_{j+1}/N_j)$, using base 10 logarithms, where N_j is the

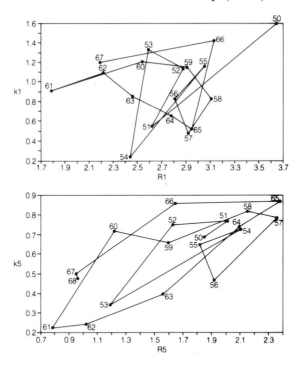

Figure 7.6   Possible cases of spiral plots of the k_1 and k_5 values for the winter moth data of Table 7.1.

number entering stage j of the life cycle in one generation. Also, let $R_j = \log(N_j)$, so that $k_j = R_j - R_{j+1}$. Furthermore, assume that the k_j values are dependent on the numbers entering stage j by a regression relationship

$$k_j = \tau_j + \delta_j R_j + \varepsilon_j,$$

where τ_j and δ_j are constants, and the ε_j values are random independent disturbances with mean zero and variance var(ε_j).

It follows from these assumptions and definitions that

$$R_j = R_{j-1} - k_{j-1} = (1 - \delta_{j-1})R_{j-1} - \tau_{j-1} - \varepsilon_{j-1}.$$

Repeated use of this result shows that the logarithm of the number entering the final stage can be expressed as

$$R_q = (1 - \delta_1)(1 - \delta_2) \ldots (1 - \delta_{q-1})R_1 - (1 - \delta_2)(1 - \delta_3) \ldots$$
$$(1 - \delta_{q-1})(\tau_1 + \varepsilon_1)$$
$$- (1 - \delta_3)(1 - \delta_4) \ldots (1 - \delta_{q-1})(\tau_2 + \varepsilon_2) - \ldots$$
$$- (1 - \delta_{n-1})(\tau_{n-2} + \varepsilon_{n-2}) - \tau_{n-1} + \varepsilon_{n-1}.$$

More concisely, this is written as

$$R_q = \theta_0 R_1 - \sum_{j=1}^{q-1} \theta_j \tau_j - \sum_{j=1}^{q-1} \theta_j \varepsilon_j,$$

where

$$\theta_j = \begin{bmatrix} (1 - \delta_{j+1})(1 - \delta_{j+2}) \ldots (1 - \delta_{q-1}), & j = 0, 1, \ldots, q-2, \\ 1, & j = q-1. \end{bmatrix}$$

The last equation shows that the expected (mean) value of R_q is

$$E(R_q) = \theta_0 E(R_1) - \sum_{j=1}^{q-1} \theta_j \tau_j \tag{7.2}$$

and the variance is

$$\text{var}(R_q) = \theta_0^2 \text{var}(R_1) + \sum_{j=1}^{q-1} \theta_j^2 \text{var}(\varepsilon_j). \tag{7.3}$$

There are two cases to be considered for using equations (7.2) and (7.3), depending on whether or not the results for different generations are related. The unrelated case will be discussed first.

If the results for different generations are unrelated then N_1, the number entering stage 1 in any generation, can be regarded as a random variable that is independent of the number in the last stage in the previous generation. Hence, the right-hand side of equation (7.2) gives the expected value of R_q as a function of the θ and τ values, and the mean of R_1. Similarly, the right-hand side of equation (7.3) shows how the variance of R_q depends on the variance of R_1, the variances of ε_j values, and the θ parameters. Defining

$$A_u = \begin{bmatrix} \theta_0^2 \text{var}(R_1), & u = 0, \\ \theta_u^2 \text{var}(\varepsilon_u), & u = 1, 2, \ldots, q-1, \end{bmatrix} \tag{7.4}$$

it can be seen that A_0 is the contribution of the variance of R_1, and A_j $(j > 0)$ is the contribution of the variance of ε_j, to the variance of R_q. Hence, the A_j values indicate the relative importance of random variation in different stages in determining the amount of variation in the last stage. These A_j values are conveniently expressed as percentages of their sum, $\text{var}(R_q)$.

There will be times when the A_j values alone may not tell the whole story about the importance of a stage. If a good deal of the variation in the survival in a stage is a density-dependent response to the number entering the stage then the A value for the stage may be quite small because mostly the k value is removing variation introduced by earlier stages. Nevertheless, this stage will be important in the sense that if it were not present then the final stage would be more variable. In a case like this the situation will be clarified if a calculation is made to see how $\text{var}(R_q)$ can be expected to change if somehow the survival in the stage were made constant. This amounts to fixing the corresponding k value at its mean or, in other words, making the δ and ε values equal zero for the stage. From equation (7.3) it can be shown that if this is done for stage u then the value of $\text{var}(R_q)$ is expected to change to

$$B_u = \sum_{j=0}^{u-1} A_j/(1 - \delta_j)^2 + \sum_{j=u+1}^{q-1} A_j. \tag{7.5}$$

Thus, if B_u is very different from the observed variance of R_n then variation in the survival in stage u seems to be important for population dynamics. Like the A_j values, it is convenient to express the B_j values as percentages of $\text{var}(R_q)$.

Consider now the situation where results are related from one generation to the next. In particular, assume that the numbers entering stage 1 in generation i are determined by multiplying the number entering the last stage in generation $i-1$ by a constant C. Typically, this constant would be the maximum or the average number of eggs produced per adult. Thus if N_{ij} is the number entering stage j in generation i then $N_{i1} = CN_{i-1q}$, so that

$$R_{i1} = \log(N_{i1}) = \log(N_{i-1q}) + \log(C) = R_{i-1q} + \log(C).$$

From equation (7.2) this means that the relationship between the mean values of R_q in generations $i-1$ and i is

$$E(R_{iq}) = \theta_0\{E(R_{i-1q}) + \log(C)\} - \sum_{j=1}^{q-1} \theta_j \tau_j. \tag{7.6}$$

This shows that the stable value of the mean, with $E(R_{iq}) = E(R_{i-1q})$, occurs when this mean equals

$$E(R_q) = \left\{\theta_0 \log(C) - \sum_{j=1}^{q-1} \theta_j \tau_j\right\}/\{1 - \theta_0\}, \tag{7.7}$$

providing that θ_0 does not equal 1.

As far as variation is concerned, equation (7.3) shows that the change in the variation of R_q from one generation to the next is given by

$$\text{var}(R_{iq}) = \theta_0^2 \text{var}(R_{i-1q}) + \sum_{j=1}^{q-1} A_j, \tag{7.8}$$

where A_j is given by equation (7.4). Here, $\text{var}(R_{i1}) = \text{var}(R_{i-1q})$ as R_{i1} and

R_{i-1q} only differ by the constant $\log(C)$. It follows that if the variance is stable so that $\text{var}(R_{iq}) = \text{var}(R_{i-1q}) = \text{var}(R_q)$ then

$$\text{var}(R_{iq})\{1 - \theta_0^2\} = \sum_{j=1}^{q-1} A_j, \tag{7.9}$$

which is only possible if $\theta_0^2 < 1$ since both sides of the equation must be positive. If it can occur, the stable variance is

$$\text{var}(R_{iq}) = \sum_{j=1}^{q-1} A_j/(1 - \theta_0^2). \tag{7.10}$$

On the other hand, if $\theta_0^2 \geq 1$ then the population will be undergoing an 'explosion' and the whole concept of key factor analysis becomes meaningless.

If $\theta_0^2 < 1$ then equations (7.9) and (7.10) show that the relative importance of random variation in stage u (in comparison to the other stages) in determining the long-term variance of R_q depends on the value of A_u as a percentage of ΣA_j. Also, if the survival in stage u has a strong influence on the variation in the final stage because it reduces the variation produced in earlier stages then this should show up in ΣA_j, changing substantially when the survival rate in the stage is made constant by changing δ_u and $\text{var}(\varepsilon_u)$ to zero. The expected change to ΣA_j when this is done is given by equation (7.5) with $A_0 = 0$. It is convenient to express this change as a percentage of ΣA_j.

In order to use equations (7.4) to (7.10) to detect key factors it is necessary to estimate the δ and $\text{var}(\varepsilon)$ values. This can be done for each stage using standard linear regression methods. For G generations the estimate of δ_j is then

$$\hat{\delta}_j = \sum_{i=1}^{G} (k_{ij} - \bar{k}_j)(R_{ij} - \bar{R}_j)/ \sum_{i=1}^{G} (R_{ij} - \bar{R}_j)^2,$$

where k_{ij} and R_{ij} are values for stage j in generation i, and $\bar{k}_j$ and $\bar{R}_j$ are mean values for the same stage. The estimate of τ_j is

$$\hat{\tau}_j = \bar{k}_j - \hat{\delta}_j \bar{R}_j.$$

The estimates of $\text{var}(\varepsilon_j)$ and $\text{var}(R_j)$ have to be taken as

$$\hat{\text{var}}(\varepsilon_j) = \sum_{i=1}^{G} (k_{ij} - \hat{\tau}_j - \hat{\delta}_j R_{ij})^2/(G - 1),$$

and

$$\text{vâr}(R_j) = \sum_{i=1}^{G} (R_{ij} - \bar{R}_j)^2/(G - 1),$$

to make equation (7.3) hold for data estimates (Manly, 1979). (In regression analysis vâr(ε_j) would usually be defined with a division by $G - 2$.)

The basic assumption of the model that has been described in this section is that the k value in a stage is a linear function of R, the logarithm of the number entering that stage. There are two ways in which this assumption may be too simple. First, as mentioned in Section 7.5, the relationships between k and R values may be non-linear. The validity of the above equations will then depend on the extent to which the relationships can be approximated by linear functions. Second, a k value may depend on densities in other stages and in previous generations. Models of this type have been discussed by Royama (1981).

Example 7.5   Further Analysis of the Winter Moth Data

As has been noted before, the generations are related for the winter moth data. The number entering stage 1 in one generation is the number entering stage 7 in the previous generation multiplied by 56.25. Estimates from the equations (7.6) to (7.10) for related data are shown in Table 7.5. The sum of the A_j values is 0.084. The major contribution is A_1 at 0.053 (63.2% of the total), which suggests that k_1 is the key factor. The other A_j values suggest that k_2, k_3 and k_4 are not at all important whereas k_5 and k_6 make a moderate contribution. The B_1 value confirms the importance of k_1. If the survival in this stage became constant then ΣA_j would apparently decrease to about 0.031 (36.8% of the present value). Making k_2 to k_4 constant would have very little effect whereas making k_6 constant would make a minor change.

Table 7.5   Estimates of parameters for a key factor analysis of the winter moth data in Table 7.1

J	τ	δ	var(ε)	θ	A	A%	B	B%	var(R)
0	–	–	–	0.553	–	–	–	–	–
1	0.531	0.161	0.130	0.659	0.053	63.2	0.031	36.8	0.184
2	−0.011	0.021	0.001	0.673	0.000	0.5	0.086	102.2	0.252
3	0.062	−0.019	0.000	0.661	0.000	0.2	0.082	97.5	0.243
4	0.020	0.006	0.000	0.665	0.000	0.1	0.085	100.7	0.252
5	0.102	0.313	0.019	0.967	0.017	19.8	0.128	151.8	0.249
6	0.116	0.033	0.015	1.000	0.014	16.2	0.076	89.5	0.136
7	–	–	–	–	–	–	–	–	0.141

However, k_5 seems to be having a strong stabilizing effect because making this constant would increase ΣA_j by about 51.8%.

Because the absolute value of θ_0 is less than 1, the population seems stable and equations (7.7) and (7.10) can be used to predict the long-term mean and variance of R_7. From equation (7.7) it is calculated that $E(R_7) = 0.797$, which corresponds to the mean number entering the stage being approximately $10^{0.797} = 6.3$. This seems realistic considering the observed numbers shown in Table 7.1. From equation (7.10) it is calculated that the stable variance of R_7 is 0.1214. This also seems a realistic value as the observed variance of R_7 for the data is 0.1405. Generally, this analysis confirms the graphical Varley and Gradwell analysis and also the Podoler–Rogers–Smith regression-based analysis.

Example 7.6   Further Analysis of the Tawny Owl Data of Table 7.2

The tawny owl data of Table 7.2 are an example of a situation where the numbers entering stage 1 in each generation are not directly related to the numbers in the final stage in the previous generation. This means that the variance of R_1 needs to be considered as a potential key factor. Estimates for the analysis are shown in Table 7.6. Here, the key factor seems to be k_5 since it contributes most to the variance of R_6, and this variance would be increased considerably if k_5 were made constant. The second most important factor is k_1. The other k values seem rather unimportant. One thing that this analysis picks up that has been missed by the previous analyses is the fact that the variation in k_1 is apparently to a large extent a density-dependent reaction to variation in the numbers entering stage 1. This is the reason why the key factor is found to be k_5 rather than k_1.

7.7   THE RELATIVE MERITS OF DIFFERENT METHODS OF KEY FACTOR ANALYSIS

Three methods for carrying out a key factor analysis have been described and used on the winter moth and tawny owl data of Tables 7.1 and 7.2: the graphical method of Varley and Gradwell (1960); the regression method of Podoler and Rogers (1975), with an extension suggested by the work of Smith (1973); and the method of Manly (1977b, 1979). All methods have indicated the same key factors, but the latter two methods are helpful in clarifying the importance of correlations between k values. In practice, there is no reason why the three methods should be regarded as alternatives. It is better to think of them as being complementary, and carry out all three on all sets of data.

Smith (1973) also suggested using a principal component analysis to study the relationships between k values, and showed what the outcome was for the tawny owl data. A principal component analysis may highlight rela-

Table 7.6   Estimates of parameters for a key factor analysis of the tawny owl data of
Table 7.2

J	τ	δ	var(ε)	θ	A	A%	B	B%	var(R)
0	–	–	–	0.205	0.000	1.0	0.033	99.0	–
1	−0.863	0.567	0.061	0.474	0.012	36.2	0.023	68.2	0.008
2	0.349	−0.148	0.003	0.413	0.000	1.3	0.030	89.8	0.055
3	0.350	0.010	0.022	0.417	0.004	10.3	0.030	90.4	0.076
4	−0.102	0.090	0.006	0.458	0.001	3.3	0.036	106.8	0.094
5	−0.406	0.542	0.018	1.000	0.016	48.0	0.083	248.3	0.083
6	–	–	–	–	–	–	–	–	0.034

tionships between k values. However, it does not take into account the
known order in which these values act. As such, it is likely to be less useful
than the calculations described in the previous section.

7.8   USING SIMULATION WITH KEY FACTOR ANALYSIS

Simulation is a useful tool to determine the extent to which random variation
effects the result of key factor analysis. The most obvious use is for
determining how random variation and sampling errors are likely to effect
estimates of population parameters. This makes it possible to determine
whether estimates are statistically significant in comparison with a 'null'
model. More generally, simulation can be used as a means of assessing the
likelihood of any unusual aspect of data having arisen purely by chance.

The population model that is proposed for generating artificial key factor
data is the one described in Section 7.6, which involves the k values for
stages being linearly related to population numbers. More complicated
models that take into account delayed density-dependent mortality, the
action of predators and parasites, etc., are certainly possible. However,
these more complicated models are necessarily specific to a population in
one place in one period and, as such, do not provide a simple general method
for generating data.

Recall that $R_j = \log(N_j)$ denotes the logarithm of the number entering
stage j in a generation and $k_j = -\log(N_{j+1}/N_j)$, where

$$k_j = \tau_j + \delta_j R_j + \varepsilon_j, \qquad (7.11)$$

where τ_j and δ_j are constants and ε_j is a random 'disturbance' to the system,
with a mean value of zero. For simulation purposes, the ε_j values will be
taken to be independent, normally distributed values with constant
variances from generation to generation.

Simulations depend on whether generations are related or not. With related generations the number entering stage 1 in generation i is equal to the number entering stage q in generation $i - 1$ multiplied by a constant C. A set of population frequencies can therefore be determined by choosing an initial number N_1 entering stage 1 in generation 1; generating $k_1, k_2, \ldots,$ k_{q-1} using equation (7.11) with random ε_j values, and hence determining $R_2, R_3, \ldots, R_q$; finding R_1 in generation 2 as $R_q + \log(C)$; generating k values for generation 2 using equation (7.11), and hence determining R values, etc. In this way, simulated data can be produced for as many generations as is required.

When the data from successive generations are independent, the same scheme can be used to determine population frequencies except that the $R_1 = \log(N_1)$ value for each generation is a random value from a normal distribution with an appropriate mean and variance. However, in certain cases (such as with the tawny owl, where there is a distinct trend in R_1 values) it may be more appropriate to use the observed R_1 values as the starting-points for generations and hence make the simulated data conditional on these.

Most real data sets consist of population frequencies with superimposed sampling errors. These errors may be attached either to the population frequencies, to estimated survival rates or to a combination of both. If the nature of these errors is understood, and estimates of variances are available, then it will be a simple matter to include the errors in the simulation.

In analysing real data there are two types of simulation that will produce useful results. A null model can be used for which δ_j values are zero, τ_j values are equal to observed mean k_j values, and the variances of ε_j values are set equal to the observed variances of k_j values. The distribution of estimates produced by simulations with this model will indicate whether the estimated δ_j values for the real data are such as could easily have arisen by chance in a population with no density-dependent mortality. In other words, a test for statistically significant density-dependence becomes possible. It is also possible to check whether the B_j values for the real data provide evidence of 'hidden' key factors. The key factors indicated by A_j values may or may not be the same for the real data and the null model.

The alternative to using the null model will usually be the model with the τ_j and δ_j values, and the variance of ε_j being set equal to the estimates from the real data. The distribution of estimates produced from simulations with this model should give a good guide to the accuracy of the real data estimates in terms of biases and standard errors.

Example 7.7   Simulating Winter Moth Data

As a first example of the use of simulation consider again the winter moth data shown in Table 7.1. Estimates of population parameters for the model of Section 7.6 are given in Table 7.5. In summary, it can be said that if the model is accepted then: (1) because θ_0^2 is less than 1, the population is stable in the sense that there is a fixed long-term variance for R_7, the logarithm of the density in the adult stage; (2) variation in k_1 accounts for about 63% of ΣA_j and is therefore the key factor, followed by k_5 (20%) and k_6 (16%); (3) variation in k_5 has a stabilizing influence on the population because ΣA_j would increase by about 52% if k_5 was made constant. Furthermore, equation (7.6) predicts the long-term variance of R_7 to be 0.121.

To determine the reliability of these conclusions, taking into account the effects of random variation in k values, the null model and the estimated model have been simulated. For both models 500 independent realizations of 19 generations of the winter moth population model were produced starting with the observed number in stage 1, generation 1, of 4365 individuals per square metre. All artificial sets of data were then analysed in exactly the same way as the real data.

The most important results from the null model simulations are shown in Table 7.7(a). These are the percentages of the 500 generated populations for which an estimated parameter was larger than the estimate found with the real data. We see, for example, that 92.6% of the estimated θ_0 values were larger than the real data estimate of 0.55. Because small values indicate population stability, this indicates that the stability level estimated for the real data is not very likely to occur with data from a model that is unstable, but it is not significantly small at the 5% level. Also, the simulated values of B_5 as a percentage of ΣA_j are all less than the real data value, and only 1% of the simulated τ_5 values exceed the real data value. It seems that the apparent stabilizing effect of k_5 is real. The data value of A_1 as a percentage of ΣA_j is exceeded by 62.4% of the simulated values, but this only reflects the fact that k_1 is the key factor for the null model.

The simulations with τ_j, δ_j and var(ε_j) values equal to the estimates from the real data produced the estimates of biases and standard errors shown in parts (b) and (c) of Table 7.7. These indicate that the A_j and B_j are estimated with moderate accuracy, although they may have some minor biases. However, the estimates of τ_1, and δ_1 are strongly biased. Also, θ_0 is biased in the direction of indicating more stability than really exists.

A further idea of the reliability of the key factor analysis on the real data is given by knowing that for 97.2% of the simulated sets of data A_1 was the largest A_j value, A_5 was largest for 1.8% of sets, and A_6 was largest for the remaining 1.0% of sets. Also, for 97.6% of the simulated sets of data, B_5 was the largest B_j value, for 1.4% B_6 was largest, and for 0.6% B_4 was largest. It seems fair enough to regard this as an indication of the extent to which chance may affect the choice of key factors.

The performance of the Podoler–Rogers–Smith regression method for choosing key factors is indicated by the results in Table 7.8. It can be seen that k_1 was correctly chosen as the key factor for 99.2% of simulated sets of

Table 7.7 Simulation results for the winter moth population

(a) Percentages of simulated sets of data producing estimates greater than the estimates for the observed data (500 simulations with the null model).

j	τ	δ	θ	A	$A\%$	B	$B\%$
0	–	–	92.6	–	–	–	–
1	59.0	36.6	98.8	91.0	62.4	93.2	37.6
2	94.2	2.6	98.6	75.6	23.8	96.0	3.0
3	3.6	99.2	98.8	96.6	72.6	96.8	99.0
4	81.6	14.8	98.8	92.0	50.8	96.2	15.2
5	94.8	1.0	74.6	96.6	72.0	41.4	0.0
6	71.0	25.4	–	20.6	2.4	97.0	78.2

(b) Results from 500 simulated sets of data when the simulation model has parameter values equal to the estimates obtained from the real data: biases (simulation means minus values used in generating data).

j	τ	δ	θ	A	$A\%$	B	$B\%$
0	–	–	−0.05	–	–	–	–
1	−0.20	0.07	−0.01	0.000	−0.1	−0.003	−0.1
2	0.01	−0.01	−0.01	0.000	−0.1	−0.004	−0.4
3	0.00	0.00	−0.01	0.000	0.0	−0.003	0.2
4	0.00	0.00	−0.01	0.000	0.0	−0.003	0.0
5	−0.02	0.01	−0.01	−0.001	1.1	−0.003	4.4
6	0.00	0.00	–	−0.002	−0.9	−0.001	3.9

(c) Results from 500 simulated sets of data when the simulation model has parameter values equal to the estimates obtained from the real data: standard deviations.

j	τ	δ	θ	A	$A\%$	B	$B\%$
0	–	–	0.15	–	–	–	–
1	0.57	0.21	0.09	0.023	10.9	0.007	10.9
2	0.02	0.01	0.09	0.000	0.2	0.025	1.7
3	0.01	0.01	0.09	0.000	0.1	0.024	1.0
4	0.01	0.01	0.09	0.000	0.1	0.025	1.1
5	0.12	0.07	0.08	0.006	7.6	0.042	27.5
6	0.09	0.08	–	0.004	6.6	0.024	15.7

Table 7.8 Percentages of times that different stages are chosen as the first, second, . . ., fifth key factor using the Podoler–Rogers–Smith regression method on simulated winter moth data

Key factor	Stage					
	1	*2*	*3*	*4*	*5*	*6*
1	99.2%	0.0	0.0	0.0	0.8	0.0
2	2.0	0.0	0.0	0.0	97.6*	0.4
3	0.0	0.0	0.0	0.0	0.4	99.6*
4	0.0	6.0	94.0*	0.0	0.0	0.0
5	0.0	91.6*	6.0	2.4	0.0	0.0

*Key factor for the model used in the simulation.

data, the remaining 0.8% suggesting k_5. The correct choice for the order of the remaining factors (in the sense of agreeing with what is given by the simulation parameters) has also been obtained most of the time.

To sum up, it seems from the estimates and simulations that k_1 is the key factor, k_5 is density-dependent, and the apparent stability of the population is fairly unlikely to have occurred by chance. Nevertheless, the estimation of θ_0 seems to be biased in the direction of indicating more stability than really exists, and the estimation of the density-dependent effect of k_1 seems to be biased in the direction of indicating more density-dependence than really exists.

These conclusions are essentially in agreement with those of Varley *et al.* (1973). However, den Boer (1986) has brought to light some aspects of the winter moth data that call into question whether these conclusions can really be justified. In particular, he has argued that pupal predation (k_5) increases population variation rather than reducing it. His work has led to some controversy (den Boer, 1988; Latto and Hassell, 1988; Poethke and Kirchberg, 1988) and requires some discussion here.

The most compelling piece of evidence that den Boer (1986) presented is given in the following table, which shows the effects of fixing k_5 at the mean value for the winter moth data:

	LR(larvae)	var(R_2)	LR(adults)	var(R_7)
Original data	1.56	0.260	1.34	0.140
k_5 fixed	1.39	0.176	1.03	0.128
Change	−0.17	−0.084	−0.31	−0.012

Here, $LR = \log_{10}$(highest density/lowest density) is the logarithmic range over the 19 generations of data and the variances are of $R_2 = \log$(larvae numbers) and $R_7 = \log$(adult numbers), the LR values and variances simply being alternative measures of variation. The values for k_5 fixed are obtained by recalculating population numbers with all k_5 fixed at the observed mean with other k values as for the original data, starting with the observed number in stage 1, generation 1. Because fixing k_5 has reduced all the measures of variation it is difficult to argue that this factor has a regulating influence!

Several points need to be raised when attempting to reconcile den Boer's results with the conclusions that have been reached from the analysis given earlier. To begin with, it is important to remember that the observed data are just one realization of a process that must be considerably affected by random events. It is entirely plausible that fixing k_5 will reduce variation on average, even though it does not appear to do this for the available data. Clearly, to determine average behaviour more realizations are needed, but these are not available naturally. One apparent solution to this problem involves generating data by computer simulations. Unfortunately, if this is done then the results obtained will depend entirely on the assumptions made. They will therefore not necessarily shed much light on the situation with the true population.

For example, if the winter moth population is simulated with k_5 set equal to its observed mean and other k values density-dependent according to equation (7.11) then the long-term values of var(R_2) and var(R_7) must be higher than they would be with k_5 density-dependent. Thus, using estimated population parameters, without fixing k_5, equation (7.10) shows that the stable value for var(R_7) is 0.1214, and the stable variance of R_2 is

$$
\begin{aligned}
\mathrm{var}(R_2) &= \mathrm{var}\{(1 - \delta_1)R_1 + \varepsilon_1\} \\
&= (1 - \delta_1)^2\mathrm{var}(R_1) + \mathrm{var}(\varepsilon_1) \\
&= (1 - \delta_1)^2\mathrm{var}(R_7) + \mathrm{var}(\varepsilon_1) \\
&= (1 - 0.161)^2 0.1214 + 0.1297 \\
&= 0.2151.
\end{aligned}
$$

The observed values var$(R_7) = 0.1405$ and var$(R_2) = 0.2519$ are in reasonable agreement with these calculated values. Setting $\delta_5 = \mathrm{var}(\varepsilon_5) = 0$ in equation (7.10) shows that making k_5 constant would result in the long-term variances of R_7 and R_2 being increased to 0.3625 and 0.3847, respectively. However, the population would still be stable because the new value of θ_0 would be 0.805, which is less than 1. Hence, it seems that the density-dependence of k_5 is not essential for regulating the population. Obviously, if this model is accepted then there is no need to carry out simulations to decide what happens if k_5 is fixed.

On the other hand, a population without any form of density-dependent mortality, that is not controlled by outside factors such as weather, must build up potential variation from generation to generation and cannot be stable. Again, it is unnecessary to carry out simulations to determine this. Looked at from this point of view, the simulations of Latto and Hassell (1988), Poethke and Kirchberg (1988) and den Boer (1988) do not tell us anything that is not obvious.

However, this does not mean that simulations are completely worthless in assessing den Boer's claim that pupal predation does not have a stabilizing effect. What can be done is to simulate populations using estimated parameter values, and see what is the effect of fixing k_5 values at their mean in exactly the same way as has been done by den Boer for the real data. This will not give the same results as simulating data with k_5 fixed, for two reasons. First, with the proposed simulations the mean of k_5 will vary with each set of data rather than being fixed at the same value for all sets. Second, the other k values are not allowed to adjust in a density-dependent way to the changes in density brought about by fixing k_5. Thus, the precise effect of den Boer's modification to data is difficult to predict, although it seems that it should increase variation on average.

There is also another reason why simulation is valuable. The model makes it possible to predict the long-term variances for R_2 and R_7 but these are not the same as the variances that can be expected for the 19 generations actually observed. At the start of the first generation the number in stage 1 was unusually large so that the observed data can be expected to show more variation than 19 generations in general. Hence, in assessing the observed variation against a model it is important to see what data the model generates when it starts from the same density in stage 1 as was actually observed.

The following table provides a summary of the results obtained with 500 simulated sets of data:

	LR(larvae)	var(R_2)	LR(adults)	var(R_7)
Mean increase	−0.08	0.010	0.39	0.117
Minimum change	−0.95	−0.192	−0.46	−0.063
Maximum change	1.17	0.350	1.55	0.644
Exceeding real data	41.2%	94.0%	92.0%	97.0%

The first row shows the mean increases in LR values and variances found by fixing k_5. The second and third rows show the minimum and maximum changes seen, and the fourth row shows the percentages of sets of data that had smaller changes than those observed for the real data (taking into account the sign).

There was a good deal of variation in the results for different sets of data. Rather surprisingly, fixing k_5 resulted in a mean decrease of 0.08 in the LR value for larvae, which is certainly not what is expected from the model. However, for adults the LR value and the variance are both increased substantially, which does agree with expectations. For variances and LR (adults) the changes for the real data are exceeded by most of the simulated values and are significantly low at about the 5% level on a one-sided test.

Clearly, there are some aspects of the real winter moth data that do not agree very well with what is expected from the density-dependence of k_5. Den Boer has good grounds to doubt the assertion that pupal predation is the main regulator of the population, although the mechanism that inhibits the regulation that should take place is not clear. One relevant point noted by den Boer (1988) is the apparently non-random sequence of differences between observed and expected k_5 values from the regression of these values on R_5. This is related to the spiralness of the plot of k_5 values against the R_5 values (see Figure 7.6), which may reflect delayed density-dependent mortality. The question of whether or not the pattern of spiralness is significant is considered in the next section of this chapter.

Any conclusions drawn from the winter moth data must be tentative. Pupal predation may have some stabilizing effect (as it is difficult to see how a density-dependent factor can do anything else), but it is less than what is expected on the basis of the observed regression between k_5 and density (presumably because of inhibition by other non-random effects). Also, the decrease in variation found by den Boer by fixing k_5 may be exaggerated to some extent by stochastic effects.

One limitation with the simulation results mentioned above is the absence of any allowance for sampling errors. The main problem with doing this is the lack of information about the magnitude of these errors. However, what can be done is to see how sensitive the results are to errors of a moderate size. To this end all of the simulations were repeated with independent random errors being added to population frequencies. These errors were normally distributed with a mean of zero and standard deviations equal to 5% of generated frequencies, so that the 'observed' value of a population count N was usually within about 10% (two standard errors) of the true value. Adding errors in this way makes some small differences to the simulation results, but does not change conclusions.

Example 7.8   Simulating the Tawny Owl Population

The data on the tawny owl population in Wytham Wood (Table 7.2) provide an example where there is no simple relationship between successive generations. Under normal circumstances, the simplest way to simulate a population like this involves starting each generation with a random number of possible eggs, this being taken from a distribution with mean and variance

Table 7.9 Simulation results for the tawny owl data in Table 7.2

(a) Percentages of simulated sets of data producing estimates greater than the estimates for the observed data (500 simulations with the null model).

j	τ	δ	θ	A	A%	B	B%
0	–	–	56.0	68.0	69.0	45.8	31.0
1	58.6	42.6	67.8	46.0	47.0	45.6	48.8
2	46.2	53.0	67.8	59.2	59.6	51.4	47.0
3	51.8	48.8	72.0	60.0	60.8	47.8	44.2
4	88.8	22.2	69.0	17.0	10.0	49.8	31.0
5	73.8	31.0	–	32.6	35.0	39.0	33.2

(b) Results from 500 simulated sets of data when the simulation model has parameter values equal to the estimates obtained from the real data: biases (simulation means minus values used in generating data).

j	τ	δ	θ	A	A%	B	B%
0	–	–	0.12	0.002	5.2	0.002	−5.2
1	0.28	−0.14	0.10	0.003	−1.6	0.001	−1.3
2	−0.01	0.00	0.09	0.000	0.4	0.003	−1.0
3	0.02	−0.01	0.08	0.002	3.5	0.002	−2.9
4	0.08	−0.04	0.07	0.000	−1.6	0.004	−2.4
5	0.10	−0.07	–	−0.002	−6.0	−0.006	−18.1

(c) Results from 500 simulated sets of data when the simulation model has parameter values equal to the estimates obtained from the real data: standard deviations.

j	τ	δ	θ	A	A%	B	B%
0	–	–	0.48	0.004	8.0	0.018	8.0
1	1.47	0.77	0.22	0.013	17.9	0.010	17.7
2	0.15	0.09	0.19	0.001	1.2	0.016	7.1
3	0.35	0.22	0.16	0.004	8.5	0.016	18.5
4	0.07	0.06	0.17	0.001	1.4	0.020	7.5
5	0.20	0.17	–	0.007	21.5	0.035	120.7

Table 7.10   Percentages of times that different stages are chosen as the first, second, . . ., fourth key factor using the Podoler–Rogers–Smith regression method on simulated tawny owl data

Key factor	1	2	3	4	5
1	79.4*	0.0	12.8	0.0	7.8
2	13.2	0.0	26.4	0.0	60.4*
3	7.4	1.2	59.6*	0.2	31.6
4	0.0	67.4*	1.0	31.4	0.2

*Key factor for the model used in the simulation.

equal to the observed mean and variance. In the present case, there was a distinct trend from 60 possible eggs in 1949 to 110 possible eggs in 1959 due, presumably, to some environmental changes. It therefore seems best to take the numbers entering stage 1 equal to their observed values. Simulation can then be used to determine the distributions of estimates that can be expected from random variation in k values only. Recall from the analyses of these data given earlier in this chapter it seems that k_5 is the key factor, with k_1 acting to reduce variation due to the numbers entering stage 1.

Five hundred simulated sets of data from the null model, with the observed R_1 value for each generation, were generated to determine the significance of the estimates shown in Table 7.6. Table 7.9(a) shows the percentages of times that the observed estimates were exceeded by simulated estimates. None of the percentages is extremely high or extremely low. Hence, the observed estimates are well within the bounds expected from a population with no density-dependent effects. This result is very easy to understand from the results shown in parts (b) and (c) of Table 7.9, which give estimated biases and standard deviations from 500 simulated sets of data with population parameters equal to the estimates of Table 7.6. The standard deviations in particular are rather large. Table 7.10 indicates the reliability of the Podoler–Rogers–Smith regression method for choosing key factors. Clearly, the choice of key factors depends on chance to a far larger extent than is desirable.

Generally, these simulations demonstrate that the data available are not sufficient to draw any firm conclusions about dynamics of this population except that variation in k_1 and k_5 is probably relatively important.

7.9   TESTING FOR DELAYED DENSITY-DEPENDENT MORTALITY

The final matter to be considered concerns the measurement of the 'spiralness' of plots like those shown in Figures 7.5 and 7.6 of k values against logarithms of the population density. Also, there is the question of how an observed amount of spiralness can be tested to determine whether it is significant in the statistical sense. As mentioned earlier, one explanation for this type of plot, if it is anticlockwise, is delayed density-dependent mortality due to a Nicholsonian parasite. Hence significant spiralness in the anticlockwise direction may be evidence of this phenomenon.

To some extent a spiralness index must be arbitrary. However, one method to calculate an index involves noting that an anticlockwise spiral or circular pattern implies that the angle θ formed by plotting three consecutive k values (as indicated on Figure 7.5) will be nearly constant, and less than 180°. One measure of spiralness is therefore the proportion of angles less than 180°. Another is the variation in θ values for successively plotted k values, measured in an appropriate way.

To formalize these ideas, let θ_i be the angle formed by plotting the k values for generations i, $i+1$ and $i+2$ for $i = 1, 2, \ldots, G-2$, where G is the number of generations of data available (Figure 7.5). Let S_1 be the proportion of these angles that are less than 180°. This is the first spiralness index.

Measuring the average and variation in a set of angles is complicated by the fact that two numerically different angles can be equivalent. For example, 0° and 360° are the same. Their average is therefore 0° (or 360°), not 180°. The standard way of overcoming this problem (Mardia, 1982) involves calculating

$$ c = \sum_{i=1}^{G-2} \cos(\theta_i)/(G-2) \qquad \text{and} \qquad s = \sum_{i=1}^{G-2} \sin(\theta_i)/(G-2) $$

and taking the mean direction as θ_0 and the variation as S_2 where $c = S_2 \cos(\theta_0)$ and $s = S_2 \sin(\theta_0)$. The second proposed spiralness index is therefore $S_2 = \sqrt{(c^2 + s^2)}$. This lies between 0 and 1, with $S_2 = 1$ indicating no variation in θ_i values (perfect spiralness) and small values indicating a good deal of variation (no spiralness).

To assess the significance of observed values of S_1 and S_2, the distributions obtained from simulated data can be considered (Manly 1989b). This will show the distributions that are liable to occur by chance alone without delayed density-dependent mortality, and the probabilities of obtaining values as large as those observed for this null hypothesis case. The latter probabilities are the 'significance' of the observed values in the statistical sense. It is appropriate to carry out the simulations using the values of τ_j, δ_j and $\text{var}(\varepsilon_j)$ estimated from the data to be tested in order to make the generated data as close as possible to the real data except for any possible delayed density-dependent effects.

Berryman (1981, 1986, 1988) has considered the situation where population densities in one stage, $N_1, N_2, \ldots, N_q$ are available for q successive generations of a population. He plots $\log(N_{t+1}/N_t)$ against N_t to produce

Table 7.11 Simulation results for testing for delayed density-dependent mortality with the winter moth data in Table 7.2

	Stage					
	1	*2*	*3*	*4*	*5*	*6*
Observed S_1	0.77	0.47	0.59	0.65	0.88	0.35
Simulation mean	0.70	0.51	0.51	0.51	0.57	0.56
Simulation SD	0.11	0.13	0.14	0.13	0.13	0.14
Significance level	21.8%	53.6%	20.6%	10.0%	0.0%	89.0%
Observed S_2	0.45	0.12	0.12	0.07	0.47	0.15
Simulation mean	0.43	0.22	0.22	0.22	0.30	0.31
Simulation SD	0.13	0.15	0.15	0.15	0.14	0.14
Significance level	46.8%	70.4%	68.4%	87.4%	12.2%	86.4%

what are called 'phase portraits'. A cyclic plot is then evidence of a delayed feedback. He suggests that the length of the lag can be determined by plotting $r_t = \log(N_{t+1}/N_t)$ against N_{t-h}, for $h = 1, 2, 3$, and so on. The appropriate lag is then the one that most clearly compresses the data to a single line (Royama, 1977; Berryman, 1981, p. 64).

Phase portraits are not quite the same as plots of k values against R values because k values refer to mortality through a single stage and r values refer to changes through a whole generation. However, there seems no reason why k values should not be plotted against R values lagged by more than one generation to give a graphical analysis analogous to Berryman's phase portrait analysis.

Example 7.9   Testing the Winter Moth Data for Delayed Density-Dependence

The values of S_1 and S_2 for the winter moth data are shown in Table 7.11, together with simulation means and standard deviations, and significance levels. The simulation results are for 500 sets of data. The observed S_1 values are all within the bounds expected from the null model except for stage 5. In that case, the data value is larger than all of the simulated values. There is clear evidence here of a significant anticlockwise cycle, possibly related to delayed density-dependent mortality. This is not picked up by the S_2 statistic for which there are no significant results in any stage.

7.10   RECENT DEVELOPMENTS

Hassell *et al.* (1987) have pointed out that a classical key factor analysis may not detect density-dependent processes because no account is taken of the spatial spread of a population. They give an example involving a population of the viburnum whitefly *Aleurotrachelus jelinekii* on a single bush. With total counts in stages analysed for 16 generations there is no evidence of density-dependent mortality. However, when the results on 30 labelled leaves are considered, evidence is found for density-dependent mortality in eight out of nine generations.

Liebhold and Elkinton (1989) have suggested that a more informative analysis is possible when data are collected at points over a spatial grid, to give a matrix of life tables for each generation of a population. They give an example of this type of study on the gypsy moth *Lymantria dispar* on Otis Airbase, Massachusetts, from 1985 to 1987.

Bellows *et al.* (1989) have addressed the problem of quantifying the impact of parasites on insect life tables. They note that estimates of total losses due to parasitism are not usually available from stage-frequency data, and suggest two approaches for determining such estimates. The first uses direct estimates of recruitment to the susceptible host stage and to the immature parasitoid stage. The second approach uses the method of South-wood and Jepson (1962) to estimate recruitment to the host and parasitoid populations.

7.11   COMPUTER PROGRAM

The calculations for the examples in this chapter can be carried out by a computer package that is referred to in the Preface.

EXERCISE

1. Table 1.4 shows data for 12 generations of the pine looper *Bupalus piniarius* collected by Klomp (1966) for the population at Hoge Veluwe, The Netherlands. Analyse these results using the various methods described in this chapter. Use simulation to determine the accuracy and significance of parameter estimates, and to assess the effects that would result from any apparent density-dependent effects disappearing. Compare your conclusions with those of den Boer (1987). Note that the number of potential eggs in one generation is (apart from rounding errors) the number of reproducing females in the previous generation times 216.

8 Case studies

8.1 INTRODUCTION

The previous chapters have considered various general methods for obtaining and analysing data from stage-structured populations. In this final chapter, some consideration is given to some more specific approaches to modelling that can be used on particular populations by looking at five examples.

Three of the examples involve simulation (i.e., the generation of population frequencies over time using equations that determine the frequency at time $t + 1$ from the frequency at time t. This is not really surprising because simulation offers a relatively simple way to approximate the behaviour of complex populations.

Although simple simulations can be carried out on a hand calculator, most involve the use of a computer. Programs can then be written in standard languages such as BASIC, PASCAL and FORTRAN. This seems to be the approach that was used in the examples. However, there are also some simulation languages available on many computers and it may be more economical and efficient to use one of these. For example, Fargo and Woodson (1989) discuss the use of the Simulation Language for Alternative Modelling (SLAM) to model a population of the squash bug *Anasa tristis* on the squash plant *Cucurbita pepo*.

8.2 THE SHEEP BLOWFLY *LUCILIA CUPRINA*

The Australian entomologist A. J. Nicholson carried out a large number of experiments in the 1950s to examine age and density-dependent survival rates in laboratory populations of the sheep blowfly *Lucilia cuprina* (Nicholson, 1957, 1960). Here, the data from one of these populations will be considered, this being the control population kept under constant conditions for experiment L97 on the 'influence of periodic environmental changes of intrinsic oscillations'. The full data are provided by Brillinger *et al.* (1980) and much of the present discussion is based on an analysis provided by these authors. Alternative models for Nicholson's data are described by May (1976) and Gurney *et al.* (1983).

Brillinger *et al.* note that the principal stages in the life cycle of the blowfly, and their approximate durations, are as follows:

Stage	Duration
Egg	12–24 hours
Larva	5–10 days
Pupa	6–8 days
Immature adult	4 days
Mature adult	1–35 days

In the experimental population, the time from the laying of an egg to the emergence of an immature adult varied from 10 to 16 days.

The population was started on 19 May 1954 with 1000 pupae kept in a Perspex box with a grid of balsa wood on top to retain the pupal cases. The food provided consisted of lump sugar and a moistened cottonwool pad. Most flies emerged overnight. Ground liver at the rate of 0.4 g per day was added daily to the cage from 20 May onwards for the feeding of adults. The first observations on the population were made on 21 May (day 0), this consisting of counts of the number of eggs laid (E), the number of non-emerging eggs (NE), the number of emerging adults (EA), the number of dead adults (D) and the total number of adults present (T). Similar counts were taken every 2 days from then on up to day 720. Figure 8.1 shows plots of these counts, which are related by the equation

$$T_i = T_{i-1} + EA_i - D_i,$$

where the subscript indicates the observation number.

There is one important aspect of the data that is immediately apparent from Figure 8.1. This is the cycle of 35–40 days in all the series, which can be explained as follows. The initial cohort of 948 emerged flies survived well on the available food and were mature enough by day 8 to start laying eggs, which began hatching by day 20. The population then rapidly increased to a peak of about 9000 flies on day 26. By that time, the large size of the population meant that the females were not getting enough protein to produce many eggs. As a result, the population size dropped to below 100 by day 52. However, as soon as the population size dropped below about 1000 egg production increased greatly, thus setting the population up for a further increase in numbers starting on day 54. The restricted food then led to a later decrease, and the cycle of increases and decreases continued in this manner for the whole experiment. It seems that the constant food supply limiting the fecundity of the females combined with the delay between egg laying and the emergence of adults was responsible for the oscillations in the population counts.

The models for populations with continuous recruitment discussed in Section 6.8 are of no particular value with this example because both birth and death rates have been observed directly. Also, there are several reasons

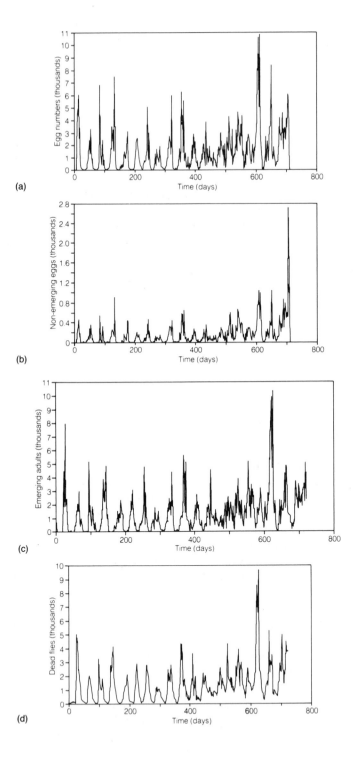

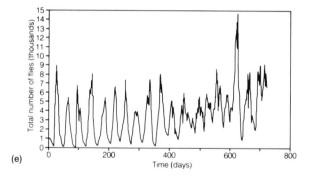

(e)

Figure 8.1 Counts of (a) eggs laid, (b) non-emerging eggs, (c) emerging adults, (d) adult deaths and (e) the total number of adults from Nicholson's (1957, 1960) control population for experiment L97, using the sheep blowfly *Lucilia cuprina*.

why the transition matrix methods for modelling reproducing populations that have been discussed in Chapter 6 cannot be used with this example. First, there is a delay of several samples between when females emerge from eggs and the time when they start laying, even when there is ample food. Second, there is also a delay of several samples between when eggs are laid and when adult flies start to emerge. These two delays mean that transitions between the egg stage and the fly stage, and between the fly stage and the egg stage are sometimes impossible between two sample times. Counting numbers in larvae, pupae and immature adult stages might overcome this problem. Third, most eggs were laid when the population size was small, which is the opposite of what the transition matrix models of Chapter 6 suggest should happen. Fourth, there is an apparent increase in the adult death rate during periods of high numbers. This is shown in Figure 8.2, which gives a plot of the death rate (dead flies in one sample divided by the number of flies in the previous sample) against the number of flies at risk (the number in the previous sample).

Brillinger *et al.* concentrated their analysis on the relationship between the number of deaths and the population sizes. They fitted two models for this relationship, using a variation of the Kalman filter and maximum likelihood. The first was an additive model with the probability of an adult aged i days dying between two sample times having the form $q_i = \alpha_i + \beta N + \gamma N_-$, where N stands for the population size at the start of the time interval, and N_- stands for the population size at the start of the previous interval. The second model was multiplicative with $q_i = 1 - (1 - \alpha_i)(1 - \beta N)(1 - \gamma N_-)$. In both cases, $\alpha_i = \alpha$ (a constant) means that mortality is not related to age, and $\beta = 0$ and $\gamma = 0$ implies no density dependence. The multiplicative model gave a slightly better fit to the data than the additive model when 10 age ranges were used for α values.

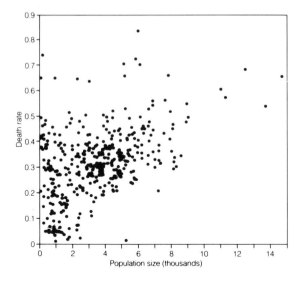

Figure 8.2   Death rates of adult blowflies plotted against the population size.

Observed and predicted deaths were in good agreement. Death rates increased markedly with age and the population size. Brillinger *et al.*'s (1980) paper should be consulted for more information about their analysis.

It is perhaps of some interest to see what results can be obtained for this example using simpler methods than were used by Brillinger *et al.* Consider death rates first. A simple model that relates these to population sizes, while restricting them to lie between zero and one, takes $q = \exp\{-\exp(\alpha + \beta N)\}$, where q is the death rate and N is the total number of adults at the start of the period when the deaths occur. Writing $\log\{-\log(q)\} = \alpha + \beta N$, this relationship can by fitted by a linear regression of $\log\{-\log(q)\}$ on N. This gives the fitted function

$$q = \exp\{-\exp(0.420 - 0.000078N)\},$$

which accounts for 19.1% of the variation in the data values of $\log\{-\log(q)\}$. The estimated standard error associated with the coefficient of N is 8.5×10^{-6}. A better result is obtained by following Brillinger *et al.* and relating the death rate to the number of adults in the previous census as well as the current number. This suggests fitting the bivariate regression $\log\{-\log(q)\} = \alpha + \beta N + \gamma N_-$, where N_- is the earlier adult count. The fitted function is then

$$q = \exp\{-\exp(0.469 + 0.000032N - 0.000125N_-)\}, \qquad (8.1)$$

which accounts for 29.5% of the variation in the data. Standard errors of the coefficients of N and N_- are both 1.7×10^{-5}. Figure 8.3 shows how the

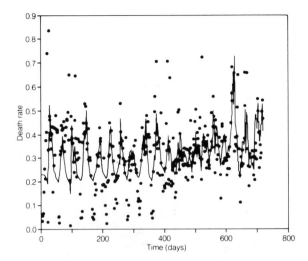

Figure 8.3   The observed death rates of adult blowflies compared with the values predicted from equation (8.1). •, observed; —, expected.

observed death rates compare with the values given by this equation. There is clearly a good deal of variation not accounted for, but the fitted function does follow the oscillations quite well.

Consider next a simple model for the number of emerging adults. This should clearly depend on the number of eggs previously laid, and the data indicate that the time between laying and emergence is at least 12 days (6 samples). This suggests an equation of the form

$$EA_i = \alpha_1 E_{i-6} + \alpha_2 E_{i-7} + \alpha_3 E_{i-8} + \ldots ,$$

going back as far as necessary with the egg laying. In fact, there seems little point in including more than three egg-laying numbers on the right-hand side of this equation. Using multiple regression, the estimated equation with three terms is

$$EA_i = 0.677E_{i-6} + 0.151E_{i-7} + 0.045E_{i-8}. \qquad (8.2)$$
$$(0.025) \qquad (0.027) \qquad (0.026)$$

This accounts for 81.5% of the variation in the number of emerging adults. Estimated standard errors are shown in parenthesis below the corresponding estimates. Figure 8.4 shows good agreement between the observed and fitted values.

The final aspect of the population that can be modelled is the fecundity of the adults. As has already been noted, this seems to have been limited by

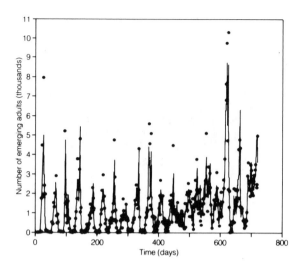

Figure 8.4 The observed emergencies of adult blowflies compared with the values predicted from equation (8.2). ●, observed; —, expected.

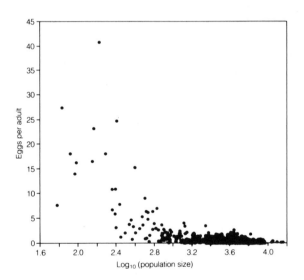

Figure 8.5 Blowfly eggs produced per adult plotted against the logarithm (to base 10) of the population size.

food. Indeed, as is shown in Figure 8.5, the number of eggs produced per adult dropped to almost zero as soon as the population size increased beyond about 1000 (so that the logarithm of size was more than 3). Because the adults cannot lay eggs until about 8 days after they have emerged, this figure does not tell the whole story. An appropriate model involves taking into account this delay, the survival of adults from the time of emergence, and the inhibiting effect that the total population size has on fecundity. This will not be pursued further here.

8.3   THE NEMATODE *PARATRICHODORUS MINOR*

Schneider and Ferris (1986) and Schneider (1989) addressed the problem of modelling the life cycle of the soil-inhabiting ectoparasitic nematode *Paratrichodorus minor*. They discuss some difficulties in gathering data caused by the fact that individuals cannot be counted directly. Instead, soil habitats must be destructively sampled and target organisms extracted from the soil. This results in greater variation than simple counting as extraction may result in some deaths, but not necessarily at the same rate in each stage. Also, the extraction process may not be equally efficient in each stage. Further, if the extraction process is not rapid enough then nematodes may change stages while this takes place.

Schneider and Ferris considered a six-stage model for *P. minor* consisting of eggs, four juvenile stages (J1–J4) and female adults. Males are rare, and were not considered. To generate data for estimating population parameters they inoculated plants with an approximately uniform cohort of J1 individuals and then destructively sampled for a period of 378 degree days, ensuring that the time between samples was less than the duration of the shortest stage. The total sampling period included several generations. The numbers extracted in each stage had to be adjusted for extraction efficiencies because it was clear that there were too many individuals in the J2 stage compared to the J1 stage. The extraction efficiencies were estimated by seeding samples with a population with a known size and age structure, and then immediately extracting the individuals. The efficiencies were then estimated for each stage as the ratio of the number recovered to the initial number (McDonald and Manly, 1989).

To model their data, Schneider and Ferris assumed that the duration of stages have Erlangian distributions. This means that stage i can be thought of as consisting of k_i substages, where the time spent in each substage has a negative exponential distribution. Development through stages can then be determined by passing individuals through substages. Mortality can be imposed at the end of each substage. Given a mean, variance and survival rate for each stage, a fecundity rate for the adult stage, and the starting age structure, Schneider and Ferris's model can be used to determine the numbers in different stages at any later time, measured in degree days. The

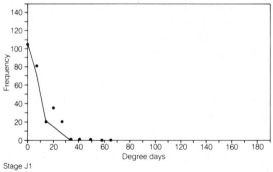

Stage J1

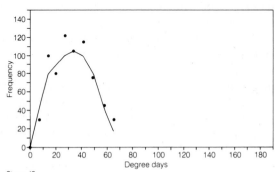

Stage J2

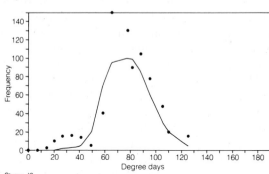

Stage J3

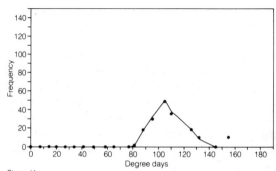

Stage J4

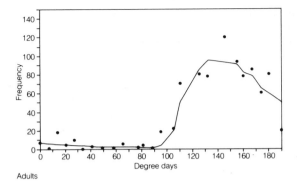

Figure 8.6 Comparison of observed and expected stage-frequencies for one generation of the nematode *Paratrichodorus minor*. Where ●, observed; —, expected.

approach of these authors to estimating parameters consists of generating populations using different combinations of parameters and finding which combination gives the smallest goodness-of-fit statistic $\Sigma(O - E)^2/O$, where O indicates an observed and E indicates an expected stage-frequency. This method of estimation is very computer intensive. For example, if four values are tested for each of three parameters for four stages then this involves generating $4^{12} = 16\ 777\ 216$ populations to find which is closest to the observed one. It seems likely that a standard procedure for function maximization such as the MAXLIK algorithm (Section 3.3) will be a much more efficient means of estimation.

With their nematode data, Schneider and Ferris began their analysis by considering only the first generation. After going through a process of coarsely fitting parameters and then fine-tuning, they obtained the original parameter estimates that are shown in Table 8.1. The data are fitted reasonably well (Figure 8.6). They then extended the model to predict numbers for the whole data set for the 378 degree days using a previously determined fecundity rate of 0.784 progeny per female per degree day, and an egg stage of 53 ± 7.4 degree days. They found at that point that the second generation J1 individuals were entering far too early in the model so a preovipositional adult period with a mean of 79 degree days was built in. Model fitting then led to the modified estimates shown in Table 8.1, some of which are rather different from the original estimates. The fit is fairly reasonable for the whole data set, but with some apparently systematic deviations between observed and expected frequencies (Figure 8.7). For example, predicted numbers in J1 are too high, and then too low from about 240 degree days on. These systematic deviations may, of course, be a result of population parameters changing with time.

Table 8.1 Estimates of parameter values for the nematode *Paratrichodorus minor* with time measured in degree days above 10°C

Stage	Mean		Standard deviation		Survivorship	
	Original	Modified	Original	Modified	Original	Modified
Egg	53	53	7.4	7.4	1.00	1.00
J1	22	14	6.7	4.2	1.00	0.90
J2	44	50	4.4	7.5	1.00	0.87
J3	47	45	4.7	6.7	1.00	0.90
J4	14	8	1.4	3.2	1.00	0.90
Pre-adult	–	79	–	23.7	–	1.00
Adult	78	100	39.0	70.0	1.00	0.50

Schneider and Ferris's modelling and estimation procedure is flexible, and capable of finding a good fitting model in cases where more standard iterative estimation procedures such as MAXLIK do not converge. However, as has already been noted, these more standard procedures will be much more efficient when they do work. A major problem with the Schneider and Ferris procedure is that the determination of the accuracy of estimates is difficult. Either experimental replication or simulation can be used to determine the amount of variation to be expected in data. However, to find the variation that this generates in parameter estimates requires the computer-intensive fitting process to be used many times.

8.4 THE PINK COTTON BOLLWORM MOTH *PECTINOPHORA GOSSYPIELLA*

It was mentioned in Section 8.2 that Brillinger *et al.* (1980) used the Kalman filter and maximum likelihood estimation in modelling the relationship between the number of deaths and the population size of the sheep blowfly. Blough (1989) has also noted the value of this approach, and used it with a study of the pink cotton bollworm moth *Pectinophora gossypiella* in a southern Arizona cotton field.

Data were collected on the number of trapped moths and the amount of irrigation water present, at three locations, for 103 days from 27 May to 26 September, 1986. Recordings were made daily except on days 13–21, 36–45, 57–60 and 81. Insecticide was applied on days 53, 58, 65, 80 and 87 so no moth counts were made on those days. This resulted in two-day aggregate observations being taken on days 54, 66 and 88. Day 82 was a three-day aggregate and day 72 was a two-day aggregate. Blough was interested in using the data to relate moth numbers at the three sampling locations to the numbers on the previous day, the amount of irrigation water and the applications of insecticide. The model used was

$$\log(\mathbf{y}_t + \mathbf{1}) = \log(\mathbf{y}_{t-1} + \mathbf{1}) = \mathbf{A}\mathbf{z}_t + \mathbf{B}\mathbf{e}_{t-1} + \mathbf{e}_t,$$

where $\mathbf{y}_t$ is a 3×1 vector of moth counts on day t, $\mathbf{1}$ is a 3×1 vector of 1's, $\mathbf{A}$ is a 3×3 matrix of regression coefficients, $\mathbf{z}_t$ is a 3×1 vector of amounts of irrigation water, $\mathbf{B}$ is a spatial weight matrix and $\mathbf{e}_t$ is a vector of random effects. Interventions are implicitly included in the state–space formulation of the model, as discussed by Blough.

From his analysis, Blough concluded that the application of insecticide had a significant effect on population numbers on days 53 and 65. No significant effect was found for the other three applications. Also, there was no significant effect for the amounts of irrigation water.

Blough's approach to modelling seems to have considerable potential use, although the stage structure of the population is not accounted for with his example. Missing values, aggregate data, transformations, covariates, interventions and spatial data can all be handled in a fairly straightforward way. His paper should be consulted for more details.

8.5 THE SOUTHERN PINE BEETLE *DENDROCTONUS FRONTALIS*

The southern pine beetle is a major pest of pine forests in the southern United States, causing more than 100 million dollars' worth of damage in 1986 alone (Connor *et al.*, 1987, quoted by Lih and Stephen, 1989). The possibility of a major infestation is a constant concern to forest managers. Outbreaks usually originate in spring or summer, a mass attack on trees being triggered by pheromones of the beetles and odours of the host trees. The resin system of a tree is its main defence as it can flush beetles from the tree or trap them. However, this can be overcome if enough beetles attack in a short period of time. The beetles then bore through the bark to the phloem/cambium interface, mate, and construct tunnels with egg niches on both sides. Parent beetles may then re-emerge from the tree and attack another neighbouring tree. Young larvae emerge from eggs and begin feeding on the phloem/cambium tissue of the tree. Later, they enter the outer bark to pupate and to complete development. Emerging brood adults then join other adult beetles in attacking attractive trees at what is called the 'active front' of the 'spot' of infestation. In this way, infestations continue to increase in size as the season develops. From three to nine overlapping generations develop each year, depending on the location. Mortality occurs in all stages.

Taha and Stephen (1984) and Lih and Stephen (1989) describe a computer simulation model SPBMODEL that is designed to predict the growth of infestation over a 3-month period for pine stands that are already infested

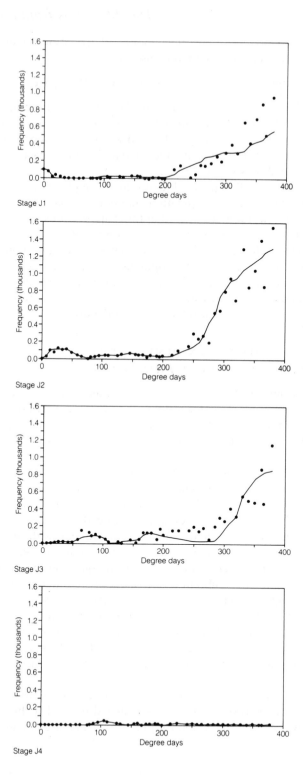

Stage J1

Stage J2

Stage J3

Stage J4

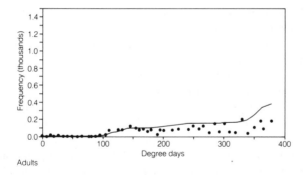

Figure 8.7 Comparison of observed and expected stage-frequencies for 378 degree days of the nematode *Paratrichodorus minor*. Where □, observed; —, expected.

with southern pine beetles. The model simulates reproduction, development and mortality of beetles within and between trees. It estimates the number of affected trees, the cumulative total of dead trees with the associated volume and dollar losses in loblolly and shortleaf pine stands. The model is intended mainly for use by managers of forest pests, who must decide whether to control a spot using salvage or cutting, or to let it run its natural course. The decision depends on a number of factors, with the probable final level of infestation being of major importance.

An iterative method is used to account for the changes in all the stages except emerging adults. With this method, which is called the rate summation approach (Wagner *et al.*, 1985), the beetle population is represented as a collection of cohorts with individuals in a cohort sharing the same developmental status. The cumulative development of each cohort is updated every hour, with rates of development, attack or production, depending on the current temperature. When the cumulative development of a cohort reaches or exceeds 1, the cohort size is reduced by the appropriate mortality factor and the remaining individuals move to the next life stage. The emerging and re-emerging adults form a pool that is reduced during the day to account for flight. When beetles in flight land on pine surfaces they become attacking adults. Flight occurs when the temperature rises above a threshold value, in the daytime only.

Model parameters have been determined in various ways. Nine years of intensive sampling has allowed regression equations to be developed for predicting within-tree mortality rates from tree and stand characteristics and the time of year. The sample data have also been used to develop equations relating attack and egg densities to stand characteristics. Laboratory rearing of eggs and brood adults at constant temperatures have been used as the basis for estimating rates of development as a function of temperature. The mortality rates of the attacking adults, the mortality rates of the in-flight beetles and the rearrival rate of the adult beetles were adjusted by matching observed data with the output from the model.

Lih and Stephen (1988) discuss the accuracy of the model in predicting up to 92 days ahead. They quote a mean absolute error of 13.3% for predictions of the cumulative number of dead trees and a mean absolute error of 52.5% for predicting the number of currently infested trees.

The model is being constantly improved to make it more accurate. Current research is concerned with learning more about the resistance and suitability of the host trees, and developing a model that expresses the stand risk to infection as a function of the size of the beetle population and the physiological condition of the trees. There are also plans to extend its use to a wider area in the United States, and for use all the year. (See Lih and Stephen's (1988) paper for more details about these developments.)

Extensive and detailed computer models such as SPBMODEL are being developed increasingly for important pest populations. Other examples are Holt *et al.*'s (1987) model of the brown planthopper *Nilaparvata lugens* on rice in the Philippines and Naranjo and Sawyer's (1989a,b) model of the northern corn rootworm *Diabrotica barberi* on field corn.

8.6   THE GREY PUP SEAL *HALICHOERUS GRYPUS*

Pup production of the grey seal *Halichoerus grypus* is most accurately determined if censuses are made throughout the whole breeding season. However, it is often convenient if an estimate can be derived from a count of pups made at just one point in time. To facilitate this, Radford *et al.* (1978) estimated population parameters using stage-frequency data obtained by the expedition to the island of North Rona off northern Scotland in 1972, and derived a method for estimating pup production from a single sample, assuming a normal distribution for birth times.

Although wild pups cannot be accurately aged, they can be assigned fairly accurately to one of five developmental stages based on physical features. Seals pass through these stages in sequence, and then leave the land. The North Rona expedition made counts of the numbers in these five stages 12 times during the breeding season. They also determined the average birth rate between counts, the average mortality rate between counts, and the average emigration rate between counts.

Radford *et al.* used the North Rona birth, death and emigration rates as input to a simulation model for pup development. The model was based on a daily transfer of pups from age i to age $i + 1$, as indicated by Figure 8.8. Births are added into day class 1 and mortality applied to that figure. The survivors are then added to day class 2, and the procedure continued for every day of the simulated season. The value of N was chosen to ensure that the known maximum stay on land can occur. Fully moulted pups are removed from day class N according to the emigration rate.

Radford *et al.* estimated the durations of stages by summing consecutive day classes for all possible durations, and finding those values for which a chi-squared statistic is smallest for the comparison of observed and expected frequencies. The best-fitting durations (6, 6, 7 and 4 days, respectively, for stages 1–4) gave a fairly good agreement between the observed and expected frequencies. Significant discrepancies were attributed to natural variability in stage durations and errors of classification made in the field. The duration of stage 5 (the time between entering the stage and emigration) was estimated at 9 days.

To estimate the pup production of a population from a single sample of stage-frequencies, Radford *et al.* suggest using the North Rona mortality

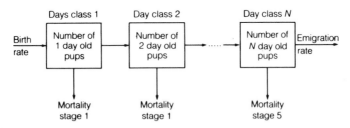

Figure 8.8   Simulation model for the grey pup seal *Halichoerus grypus*.

rates together with the stage durations estimated from their simulation model. The number observed in a stage divided by the stage duration gives an estimate of the mean birth rate of surviving pups from those born a certain number of days before. For example, stage 2 pups are assumed to be born from 7 to 12 days before the sample time. Assuming a constant birth rate during the time when the individuals in a stage were being born, it is then possible to estimate the number born each day by increasing the observed counts to allow for deaths. To estimate the full distribution of births, including pups not born at the time of the sample, it is assumed that birth dates follow a normal distribution. This is fitted to the incomplete frequency data using a technique proposed by Bhattacharya (1967).

References

Aitchison, J. and Silvey, S. D. (1957) The generalization of probit analysis to the case of multiple responses. *Biometrika* **44**: 131–40.

Aksnes, D. L. and Hoisaeter, T. (1987) Obtaining life table data from stage-frequency distributional statistics. *Limnology and Oceanography* **32**: 514–17.

Arnason, A. N. and Baniuk, L. (1980) A computer system for mark–recapture analysis of open populations. *Journal of Wildlife Management* **44**: 325–32.

Ashford, J. R., Read, K. L. Q. and Vickers, G. G. (1970) A system of stochastic models applicable to animal population dynamics. *Journal of Animal Ecology* **39**: 29–50.

Begon, F. J. (1979) *Investigating Animal Abundance*. Edward Arnold, London.

Bellows, T. S. and Birley, M. H. (1981) Estimating developmental and mortality rates and stage recruitment from insect stage-frequency data. *Researches on Population Ecology* **23**: 232–44.

Bellows, T. S., Ortiz, M., Owens, J. C. and Huddleston, E. W. (1982) A model for analysing insect stage-frequency data when mortality varies with time. *Researches on Population Ecology* **24**: 142–56.

Bellows, T. S., Van Driesche, R. G. and Elkinton, J. (1989) Life tables and parasitism: estimating parameters in joint host–parasitoid systems. In *Estimation and Analysis of Insect Populations* (eds L. L. McDonald, B. F. J. Manly, J. A. Lockwood and J. A. Logan), pp. 70–80. Springer-Verlag Lecture Notes in Statistics 55. Springer-Verlag, Berlin.

Benton, M. J. (1988) Degree days and thermal efficiency: a case against their use in describing aquatic insect growth and thermal optima. *Evolutionary Theory* **8**: 155–61.

Bernardelli, H. (1941) Population waves. *Journal of the Burma Research Society* **31**: 1–18.

Berryman, A. A. (1981) *Population Systems: A General Introduction*. Plenum Press, New York.

Berryman, A. A. (1986) On the dynamics of blackheaded budworm populations. *Canadian Entomologist* **118**: 775–9.

Berryman, A. A. (1988) The theory and classification of outbreaks. In *Insect Outbreaks* (eds P. Barbosa and J. C. Shultz), pp. 3–30. Academic Press, San Diego.

Bhattacharya, C. G. (1967) A simple method of resolving a distribution into Gaussian components. *Biometrics* **23**: 115–35.

Birley, M. (1977) The estimation of insect density and instar survivorship functions from census data. *Journal of Animal Ecology* **46**: 497–510.

Birley, M. (1979) The estimation and simulation of variable development period, with applications to the mosquito *Aedes aegypti* (L.). *Researches on Population Ecology* **21**: 68–80.

Blough, D. K. (1989) Intervention analysis in multivariate time series via the Kalman filter. In *Estimation and Analysis of Insect Populations* (eds L. L. McDonald, B. F. J. Manly, J. A. Lockwood and J. A. Logan), pp. 389–403. Springer-Verlag Lecture Notes in Statistics 55. Springer-Verlag, Berlin.

Blower, J. G., Cook, L. M. and Muggleton, J. (1981) *Estimating the Size of Animal Populations*. George Allen and Unwin, London.

Boyce, M. S. (1977) Population growth with stochastic fluctuations in the life table. *Theoretical Population Biology* **12**: 366–73.

Bradley, J. S. (1985) Comparative demography of four species of grasshoppers on a common site. In *Case Studies in Population Biology* (ed. L. M. Cook), pp. 61–100. Manchester University Press, Manchester.

Braner, M. (1988) Dormancy, dispersal and staged development: ecological and evolutionary aspects of structured populations in random environments. PhD thesis, Cornell University, New York.

Braner, M. and Hairston, N. G. (1989) From cohort data to life table parameters via stochastic modelling. In *Estimation and Analysis of Insect Populations* (eds L. L. McDonald, B. F. J. Manly, J. A. Lockwood and J. A. Logan), pp. 81–92. Springer-Verlag Lecture Notes in Statistics 55. Springer-Verlag, Berlin.

Brillinger, D. R., Guckenheimer, J., Guttorp, P. and Oster, G. (1980) Empirical modelling of population time series data: the case of age and density dependent vital rates. *Lectures on Mathematics in the Life Sciences* **13**: 65–90.

Brown, K. M. (1975) Estimation of demographic parameters from sampling data. *American Midland Naturalist* **93**: 454–9.

Brownie, C., Hines, J. E. and Nichols, J. D. (1986) Constant parameter capture–recapture models. *Biometrics* **42**: 561–74.

Burnham, K. P. (1989) Numerical survival rate estimation for capture–recapture models. In *Estimation and Analysis of Insect Populations* (eds L. L. McDonald, B. F. J. Manly, J. A. Lockwood and J. A. Logan), pp. 416–35. Springer-Verlag Lecture Notes in Statistics 55. Springer-Verlag, Berlin.

Burnham, K. P., Anderson, D. R., White, G. C., Brownie, C. and Pollock, K. (1987) *Design and Analysis for Fish Survival Experiments Based on Release–Recapture*. Monograph 5, American Fisheries Society, Bethesda, Maryland.

Carter, N., Aikman, D. P. and Dixon, A. F. G. (1978) An appraisal of Hughes' time-specific life table analysis for determining aphid reproductive and mortality rates. *Journal of Animal Ecology* **47**: 677–87.

Caswell, H. (1972) On instantaneous and finite birth rates. *Oecologia* **17**: 787–91.

Caswell, H. (1986) Life cycle models for plants. *Lectures on Mathematics in the Life Sciences* **18**: 171–233.

Caswell, H. and Twombly, S. (1989) Estimation of stage-specific demographic parameters for zooplankton populations: methods based on stage-classified matrix projection models. In *Estimation and Analysis of Insect Populations* (eds L. L. McDonald, B. F. J. Manly, J. A. Lockwood and J. A. Logan), pp. 93–107. Springer-Verlag Lecture Notes in Statistics 55. Springer-Verlag, Berlin.

Cherrill, A. J. and Begon, M. (1989) Timing of life cycles in a seasonal environment: the temperature-dependence of embryogenesis and diapause in a grasshopper (*Chothippus brunneus* Thunberg). *Oecologia* **78**: 237–41.

Clobert, J., Lebreton, J. D. and Allaine, D. (1987) A general approach to survival rate estimation by recaptures or resightings of marked birds. *Ardea* **75**: 113–42.

Cochran, W. G. (1977) *Sampling Techniques.* Wiley, New York.

Connor, M. D., Remion, M., Solomon, J. and Ward, D. (1987) Survey of damage caused by forest insects in the southeast in calendar year 1986. Southern Forest Insect Work Conference held in San Antonio, Texas, August, 1987, by the Committee on Losses Caused by Forest Insects.

Cooper, R. J. (1989) Sampling forest canopy arthropods available to birds as prey. In *Estimation and Analysis of Insect Populations* (eds L. L. McDonald, B. F. J. Manly, J. A. Lockwood and J. A. Logan), pp. 436–44. Springer-Verlag Lecture Notes in Statistics 55. Springer-Verlag, Berlin.

Cormack, R. M. (1981) Loglinear models for capture–recapture experiments in open populations. In *The Mathematical Theory of the Dynamics of Biological Populations* (eds R. W. Hiorns and D. Cooke), pp. 217–35. Academic Press, London.

Croft, B. A., Michels, M. F. and Rice, R. E. (1980) Validation of a PETE timing model for the oriental fruit moth in Michigan and central California (Lepidoptera: Olethreutidae). *The Great Lakes Entomologist* **13**: 211–17.

Crosbie, S. F. and Manly, B. F. J. (1985) Parsimonious modelling of capture–mark–recapture studies. *Biometrics* **41**: 385–99.

Crouse, D. T., Crowder, L. B. and Caswell, H. (1987) A stage-based population model for loggerhead sea turtles and implications for conservation. *Ecology* **68**: 1412–23.

Curry, G. L., Feldman, R. M. and Smith, K. C. (1978) A stochastic model of a temperature-dependent population. *Theoretical Population Biology* **13**: 197–213.

Dempster, J. P. (1956) The estimation of the number of individuals entering each stage during the development of one generation of an insect population. *Journal of Animal Ecology* **25**: 1–5.

Dempster, J. P. (1961) The analysis of data obtained by regular sampling of an insect population. *Journal of Animal Ecology* **30**: 429–32.

den Boer, P. J. (1986) Density dependence and the stabilization of animal numbers. 1. The winter moth. *Oecologia* **69**: 507–12.

den Boer, P. J. (1987) Density dependence and the stabilization of animal numbers. 2. The pine looper. *Netherlands Journal of Zoology* **37**: 220–37.

den Boer, P. J. (1988) Density dependence and the stabilization of animal numbers. 3. The winter moth reconsidered. *Oecologia* **75**: 161–8.

den Boer, P. J. and Reddingius, J. (1989) On the stabilization of animal numbers. Problems of testing. 3. Confrontation with data from the field. *Oecologia* **79**: 143–9.

Dennis, B. and Kemp, W. P. (1988) Further statistical inference methods for a stochastic model of insect phenology. *Environmental Entomology* **17**: 887–93.

Dennis, B., Kemp, W. P. and Beckwith, R. C. (1986) Stochastic model of insect phenology: estimation and testing. *Environmental Entomology* **15**: 540–6.

Derr, J. A. and Ord, K. (1979) Field estimates of insect colonization. *Journal of Animal Ecology* **48**: 521–34.

DeVries, P. G. (1979) Line intercept sampling – statistical theory, applications, and suggestions for extended use in ecological inventory. In *Sampling Biological Populations* (eds R. M. Cormack, G. P. Patil and D. S. Robson), pp. 1–70. International Cooperative Publishing House, Fairland, Maryland.

Dixon, W. J. (chief ed.) (1985) *BMDP Statistical Software*. University of California Press, Berkeley.

Dobson, A. J. (1983) *Introduction to Statistical Modelling*. Chapman and Hall, London.

Dorazio, R. M. (1986) Estimating population birth rates of zooplankton when rates of egg deposition and hatching are periodic. *Oecologia* **69**: 532–41.

Eberhardt, L. L. (1978) Transect methods for population studies. *Journal of Wildlife Management* **42**: 1–31.

Edmonson, W. T. (1960) Reproduction rates of rotifers in natural populations. *Memoire dell'Istituto de Idrobiologia* **12**: 21–77.

Edmonson, W. T. (1968) A graphical method for evaluating the use of the egg ratio for measuring birth and death rates. *Oecologia* **1**: 1–37.

Fargo, W. S. and Woodson, W. D. (1989) Potential use of an engineering-based computer simulation language (SLAM) for modelling insect systems. In *Estimation and Analysis of Insect Populations* (eds L. L. McDonald, B. F. J. Manly, J. A. Lockwood and J. A. Logan), pp. 247–55. Springer-Verlag Lecture Notes in Statistics 55. Springer-Verlag, Berlin.

Fisher, R. A. and Ford, E. B. (1947) The spread of a gene under natural conditions in a colony of the moth *Panaxia dominula* L. *Heredity* **1**: 143–74.

Gabriel, W., Taylor, B. E. and Kirsch-Prokosch, S. (1987) Cladoceran birth and death rates estimates: experimental comparisons of egg-ratio methods. *Freshwater Biology* **18**: 361–72.

Gates, C. E. (1979) Line transect and related issues. In *Sampling Biological Populations* (eds R. M. Cormack, G. P. Patil and D. S. Robson), pp. 71–154. International Cooperative Publishing House, Fairland, Maryland.

Gurney, W. S. C., Nisbet, R. M. and Lawton, J. H. (1983) The systematic formulation of tractable single-species population models incorporating age structure. *Journal of Animal Ecology* **52**: 479–95.

Hairston, N. G. and Twombly, S. (1985) Obtaining life table data from cohort analyses: a critique of current methods. *Limnology and Oceanography* **30**: 886–93.

Hairston, N. G., Braner, M. and Twombly, S. (1987) Perspective on prospective methods for obtaining life table data. *Limnology and Oceanography* **32**: 517–20.

Hart, R. C. (1987) Population dynamics and production of five crustacean zooplankters in a subtropical reservoir during years of contrasting turbidity. *Freshwater Biology* **18**: 287–318.

Hassell, M. P. (1975) Density-dependence in single-species populations. *Journal of Animal Ecology* **44**: 283–95.

Hassell, M. P., Southwood, T. R. E. and Reader, P. M. (1987) The dynamics of the viburnum whitefly (*Aleurotrachelus jelinekii*): a case study of population regulation. *Journal of Animal Ecology* **56**: 283–300.

Henderson, A. E. and Hayman, B. I. (1960) Method of analysis and the influence of fleece characteristics on unit area wool production of Romney lambs. *Australian Journal of Agricultural Research* **11**: 851–70.

Hiby, A. R. and Mullen, A. J. (1980) Simultaneous determination of fluctuating age structure and mortality from field data. *Theoretical Population Biology* **18**: 192–203.

Holt, J., Cook, A. G., Perfect, T. J. and Norton, G. A. (1987) Simulation analysis of brown planthopper (*Nilaparvata lugens*) population dynamics on rice in the Philippines. *Journal of Applied Ecology* **24**: 87–102.

Hughes, R. D. (1962) A method for estimating the effects of mortality on aphid populations. *Journal of Animal Ecology* **31**: 389–95.

Hughes, R. D. (1972) Population dynamics. In *Aphid Technology* (ed. H. F. van Emden), pp. 275–93. Academic Press, London.

Ito, Y. (1972) On the methods for determining density-dependence by means of regression. *Oecologia* **10**: 347–72.

Jackson, C. H. N. (1939) The analysis of an animal population. *Journal of Animal Ecology* **8**: 238–46.

Jolly, G. M. (1965) Explicit estimates from capture–recapture data with both death and immigration – stochastic model. *Biometrika* **52**: 225–47.

Jolly, G. M. (1979) A unified approach to mark–recapture stochastic models exemplified by a constant survival rate model. In *Sampling Biological Populations* (eds R. M. Cormack, G. P. Patil and D. S. Robson), pp. 277–82. International Cooperative Publishing House, Maryland.

Jolly, G. M. (1982) Mark–recapture models with parameters constant in time. *Biometrics* **38**: 301–21.

Keen, R. and Nassar, R. (1981) Confidence intervals for birth and death rates estimated with the egg-ratio technique for natural populations of zooplankton. *Limnology and Oceanography* **26**: 131–42.

Kemp, W. P., Dennis, B. and Beckwith, R. C. (1986) Stochastic phenology model for the western spruce budworm (Lepidoptera: Tortricidae). *Environmental Entomology* **15**: 547–54.

Kemp, W. P., Dennis, B. and Munholland, P. L. (1989) Modelling grasshopper phenology with diffusion processes. In *Estimation and Analysis of Insect Populations* (eds L. L. McDonald, B. F. J. Manly, J. A. Lockwood and J. A. Logan), pp. 118–27, Springer-Verlag Lecture Notes in Statistics 55. Springer-Verlag, Berlin.

Kempton, R. A. (1979) Statistical analysis of frequency data obtained from sampling an insect population grouped by stages. In *Statistical Distributions in Scientific Work* (eds J. K. Ord, G. P. Patil and C. Taillie), pp. 401–18. International Cooperative Publishing House, Maryland.

Kiritani, K. and Nakasuji, F. (1967) Estimation of the stage-specific survival rate in the insect population with overlapping stages. *Researches on Population Ecology* **9**: 143–52.

Klomp, H. (1966) The dynamics of a field population of the pine looper, *Bupalus piniarius* L (Lep, Geom). *Advances in Ecological Research* **3**: 207–305.

Kobayashi, S. (1968) Estimation of the individual number entering each development stage in an insect population. *Researches on Population Ecology* **10**: 40–4.

Kuno, E. (1971) Sampling error as a misleading artifact in key factor analysis. *Researches on Population Ecology* **13**: 28–45.

Lakhani, K. H. and Service, M. W. (1974) Estimating mortalities of the immature stages of *Aedes cantans* (Mg.) (Diptera, Culicidae) in a natural habitat. *Bulletin of Entomological Research* **64**: 265–76.

Latto, J. and Hassell, M. P. (1988) Do pupal predators regulate the winter moth? *Oecologia* **74**: 153–5.

Lau, C. L. (1980) Algorithm AS147: a simple series for the incomplete gamma integral. *Applied Statistics* **29**: 113–14.

Lefkovitch, L. P. (1963) Census studies on unrestricted populations of *Lasioderma serricorne* (F.) (Coleoptera: Anobiidae). *Journal of Animal Ecology* **32**: 221–31.

Lefkovitch, L. P. (1964a) The growth of restricted populations of *Lasioderma serricorne* (F.) (Coleoptera: Anobiidae). *Bulletin of Entomological Research* **55**: 87–96.

Lefkovitch, L. P. (1964b) Estimating the Malthusian parameter from census data. *Nature* **204**: 810.

Lefkovitch, L. P. (1965) The study of population growth in organisms grouped by stages. *Biometrics* **21**: 1–18.

Leslie, P. H. (1945) On the use of matrices in certain population mathematics. *Biometrika* **33**: 182–212.

Leslie, P. H. (1948) Some further notes on the use of matrices in population mathematics. *Biometrika* **35**: 213–45.

Lewis, E. G. (1942) On the generation and growth of a population. *Sankhya* **6**: 93–6.

Leibhold, A. M. and Elkinton, J. S. (1989) use of multi-dimensional life tables for studying insect population dynamics. In *Estimation and Analysis of Insect Populations* (eds L. L. McDonald, B. F. J. Manly, J. A. Lockwood and J. A. Logan), pp. 360–9. Springer-Verlag Lecture Notes in Statistics 55. Springer-Verlag, Berlin.

Lih, M. P. and Stephen, F. M. (1989) Modelling southern pine beetle (Coleoptera: Scolytidae) population dynamics: methods, results and impending challenges. In *Estimation and Analysis of Insect Populations* (eds L. L. McDonald, B. F. J. Manly, J. A. Lockwood and J. A. Logan), pp. 256–67. Springer-Verlag Lecture Notes in Statistics 55. Springer-Verlag, Berlin.

Logan, J. A., Wollkind, D. J., Hoyt, S. C. and Tanigoshi, L. K. (1976) An analytical model for description of temperature dependent rate phenomena in arthropods. *Environmental Entomology* **5**: 1133–40.

Longstaff, B. C. (1984) An extension of the Leslie matrix model to include a variable immature period. *Australian Journal of Ecology* **9**: 289–93.

Mackauer, M. and Way, M. J. (1976) *Myzus persicae* Sulz., an aphid of world importance. In *Studies in Biological Control* (ed. V. L. Delucchi), pp. 51–119. Cambridge University Press, Cambridge.

Manly, B. F. J. (1974a) Estimation of stage-specific survival rates and other parameters for insect populations passing through stages. *Oecologia* **15**: 277–85.

Manly, B. F. J. (1974b) A comparison of methods for the analysis of insect stage-frequency data. *Oecologia* **17**: 335–48.

Manly, B. F. J. (1976) Extensions to Kiritani and Nakasuji's method for the analysis of stage frequency data. *Researches on Population Ecology* **17**: 191–9.

Manly, B. F. J. (1977a) A further note on Kiritani and Nakasuji's model for stage-frequency data including comments on Tukey's jackknife technique for estimating variances. *Researches on Population Ecology* **18**: 177–86.

Manly, B. F. J. (1977b) The determination of key factors from life table data. *Oecologia* **31**: 111–17.

Manly, B. F. J. (1979) A note on key factor analysis. *Researches on Population Ecology* **21**: 30–9.

Manly, B. F. J. (1985a) *The Statistics of Natural Selection on Animal Populations*. Chapman and Hall, London.

Manly, B. F. J. (1985b) Further improvements to a method for analysing stage-frequency data. *Researches on Population Ecology* **27**: 325–32.

Manly, B. F. J. (1987) A regression method for analysing stage-frequency data when survival rates vary from stage to stage. *Researches on Population Ecology* **29**: 119–27.

Manly, B. F. J. (1989a) A review of methods for the analysis of stage-frequency data. In *Estimation and Analysis of Insect Populations* (eds L. L. McDonald, B. F. J. Manly, J. A. Lockwood and J. A. Logan), pp. 3–69. Springer-Verlag Lecture Notes in Statistics 55. Springer-Verlag, Berlin.

Manly, B. F. J. (1989b) A review of methods for key factor analysis. In *Estimation and Analysis of Insect Populations* (eds L. L. McDonald, B. F. J. Manly, J. A. Lockwood and J. A. Logan), pp. 169–89. Springer-Verlag Lecture Notes in Statistics 55. Springer-Verlag, Berlin.

Mardia, K. V. (1982) Directional distributions. *Encyclopedia of Statistical Sciences* **2**: 381–6.

May, R. M. (1976) Models for single populations. In *Theoretical Ecology: Principles and Applications* (ed. R. M. May), pp. 4–25. Blackwell, Oxford.

McCullagh, P. (1983) Statistical and scientific aspects of models for qualitative data. In *Measuring the Unmeasurable* (eds P. Nijkamp, H. Leitner and N. Wrigley), pp. 39–49. Martinus Nijhoff, Dordrecht, The Netherlands.

McCullagh, P. and Nelder, J. A. (1983) *Generalized Linear Models*. Chapman and Hall, London.

McDonald, L. L. and Manly, B. F. J. (1989) Calibration of biased sampling procedures. In *Estimation and Analysis of Insect Populations* (eds L. L. McDonald, B. F. J. Manly, J. A. Lockwood and J. A. Logan), pp. 467–83. Springer-Verlag Lecture Notes in Statistics 55. Springer-Verlag, Berlin.

Mills, N. J. (1981a) The estimation of mean duration from stage frequency data. *Oecologia* **51**: 206–11.

Mills, N. J. (1981b) The estimation of recruitment from stage frequency data. *Oecologia* **51**: 212–16.

Moloney, K. A. (1986) A generalized algorithm for determining category size. *Oecologia* **69**: 176–80.

Morris, R. F. (1957) The interpretation of mortality data in studies of population dynamics. *The Canadian Entomologist* **89**: 49–69.

Morris, R. F. (1959) Single factor analysis in population dynamics. *Ecology* **40**: 580–8.

Morrison, L. M., Brennan, L. A. and Block, W. M. (1989) Arthropod sampling methods in ornithology: goals and pitfalls. In *Estimation and Analysis of Insect Populations* (eds L. L. McDonald, B. F. J. Manly, J. A. Lockwood and J. A. Logan), pp. 484–92. Springer-Verlag Lecture Notes in Statistics 55. Springer-Verlag, Berlin.

Munholland, P. L. (1988) Statistical aspects of field studies on insect populations. PhD thesis, University of Waterloo, Ontario, Canada.

Munholland, P. L., Kalbfleisch, J. D. and Dennis, B. (1989) A stochastic model for insect life history data. In *Estimation and Analysis of Insect Populations* (eds L. L. McDonald, B. F. J. Manly, J. A. Lockwood and J. A. Logan), pp. 136–44. Springer-Verlag Lecture Notes in Statistics 55. Springer-Verlag, Berlin.

Naranjo, S. E. and Sawyer, A. J. (1989a) A simulation model of northern corn rootworm, *Diabrotica barberi* Smith and Lawrence (Coleoptera: chrysomelidae), populaton dynamics and oviposition: significance of host plant phenology. *Canadian Entomologist* **121**: 169–91.

Naranjo, S. E. and Sawyer, A. J. (1989b) Analysis of a simulation model of northern corn rootworm, *Diabrotica barberi* Smith and Lawrence (Coleoptera: chrysomelidae), dynamics in field corn, with implications for population management. *Canadian Entomologist* **121**: 193–208.

Nelder, J. A. (1975) *Glim Manual, Release 2*. Numerical Algorithms Group, Oxford, England.

Nicholson, A. J. (1933) The balance of animal populations. *Journal of Animal Ecology* **2**: 132–78.

Nicholson, A. J. (1957) The self-adjustment of populations to change. *Cold Spring Harbor Symposium on Quantitative Biology* **22**: 153–73.

Nicholson, A. J. (1960) The role of population dynamics in natural selection. In *Evolution After Darwin. 1. The Evolution of Life* (ed. S. Tax), pp. 477–521. University of Chicago Press, Chicago.

Nicholson, A. J. and Bailey, V. A. (1935) The balance of animal populations. *Proceedings of the Zoological Society of London* **3**: 551–98.

Nordheim, E. V., Hogg, D. B. and Chen, S. (1989) Leslie matrices for insect populations with overlapping generations. In *Estimation and Analysis of Insect Populations* (eds L. L. McDonald, B. F. J. Manly, J. A. Lockwood and J. A. Logan), pp. 289–98. Springer-Verlag Lecture Notes in Statistics 55. Springer-Verlag, Berlin.

Osawa, A., Shoemaker, C. A. and Stedinger, J. R. (1983) A stochastic model of balsam fir bud phenology utilizing maximum likelihood parameter estimation. *Forestry Science* **29**: 478–90.

Otis, D. L., Burnham, K. P., White, G. C. and Anderson, D. R. (1978) Statistical Inference from capture data on closed animal populations. *Wildlife Monographs* no. 62.

Pajunen, V. I. (1983) The use of physiological time in the analysis of insect stage-frequency data. *Oikos* **40**: 161–5.

Paloheimo, J. E. (1974) Calculation of instantaneous birth rate. *Limnology and Oceanography* **19**: 692–4.

Plant, R. E. (1986) A method for computing the elements of the Leslie matrix. *Biometrics* **42**: 933–9.

Podoler, H. and Rogers, D. (1975) A new method for the identification of key factors from life table data. *Journal of Animal Ecology* **44**: 85–115.

Poethke, H. J. and Kirchberg, M. (1988) On the stabilizing effect of density-dependent mortality factors. *Oecologia* **74**: 156–8.

Pollard, J. H. (1973) *Mathematical Models for the Growth of Human Populations*. Cambridge University Press, Cambridge.

Pontius, J. S., Boyer, J. E. and Deaton, M. L. (1989a) Nonparametric estimation of insect stage transition times. In *Estimation and Analysis of Insect Populations* (eds L. L. McDonald, B. F. J. Manly, J. A. Lockwood and J. A. Logan), pp. 145–55. Springer-Verlag Lecture Notes in Statistics 55. Springer-Verlag, Berlin.

Pontius, J. S., Boyer, J. E. and Deaton, M. L. (1989b) Estimation of stage transition time: application to entomological studies. *Annals of the Entomological Society of America* **82**: 135–48.

Qasrawi, H. (1966) A study of the energy flow in a natural population of the grasshopper *Chorthippus parallelus* Zett. (Prthoptera Acridae). PhD thesis, University of Exeter, UK.

Radford, P. J., Summers, C. F. and Young, K. M. (1978) A statistical procedure for estimating grey seal pup production from a single census. *Mammal Review* **8**: 35–42.

Read, K. L. Q. and Ashford, J. R. (1968) A system of models for the life cycle of a biological organism. *Biometrika* **55**: 211–21.

Reddingius, J. and den Boer, P. J. (1989) On the stabilization of animal numbers. Problems of testing. 1. Power estimates and estimation errors. *Oecologia* **78**: 1–8.

Reed, T. E. (1969) Genetic experiences with a general maximum likelihood estimation program. In *Computer Applications in Genetics* (ed. N. E. Morton), pp. 27–9. University of Hawaii Press, Honolulu.

Reed, T. E. and Schull, W. J. (1968) A general maximum likelihood estimation program. *American Journal of Human Genetics* **20**: 579–80.

Richards, O. W. and Waloff, N. (1954) Studies on the biology and population dynamics of British grasshoppers. *Anti-Locust Bulletin* **17**: 1–182.

Richards, O. W., Waloff, N. and Spradbery, J. P. (1960) The measurement of mortality in an insect population in which recruitment and mortality widely overlap. *Oikos* **11**: 306–10.

Rigler, F. H. and Cooley, J. M. (1974) The use of field data to derive population statistics of multivoltine copepods. *Limnology and Oceanography* **19**: 636–55.

Royama, T. (1977) Population persistence and density-dependence. *Ecological Monographs* **47**: 1–35.

Royama, T. (1981) Fundamental concepts and methodology for the analysis of animal population dynamics, with particular reference to univoltine species. *Ecological Monographs* **51**: 473–93.

Ruesink, W. G. (1975) Estimating time-varying survival of arthropod life stages from population density. *Ecology* **56**: 244–7.

SAS Institute (1985) *SAS User's Guide: Statistics.* SAS Institute, Cary, North Carolina.

Saunders, J. F. and Lewis, W. M. (1987) A perspective on the use of cohort analysis to obtain demographic data for copepods. *Limnology and Oceanography* **32**: 511–13.

Sawyer, A. J. and Haynes, D. L. (1984) On the nature of errors involved in estimating stage-specific survival rates by Southwood's method for a population with overlapping stages. *Researches on Population Ecology* **26**: 331–51.

Schaalje, G. B. and van der Vaart, H. R. (1989) Relationships among recent models for insect population dynamics with variable rates of development. In *Estimation and Analysis of Insect Populations* (eds L. L. McDonald, B. F. J. Manly, J. A. Lockwood and J. A. Logan), pp. 299–312. Springer-Verlag Lecture Notes in Statistics 55. Springer-Verlag, Berlin.

Schneider, S. M. (1989) Problem associated with life cycle studies of a soil-inhabiting organism. In *Estimation and Analysis of Insect Populations* (eds L. L. McDonald, B. F. J. Manly, J. A. Lockwood and J. A. Logan), pp. 156–66. Springer-Verlag Lectures Notes in Statistics 55. Springer-Verlag, Berlin.

Schneider, S. M. and Ferris, H. (1986) Estimation of stage-specific development times and survivorship from stage-frequency data. *Researches on Population Ecology* **28**: 267–80.

Seber, G. A. F. (1965) A note on the multiple-recapture census. *Biometrika* **52**: 249–59.

Seber, G. A. F. (1982) *Estimation of Animal Abundance and Related Parameters*, 2nd edn. Griffin, London.

Seitz, A. (1979) On the calculation of birth rates and death rates in fluctuating populations with continuous recruitment. *Oecologia* **41**: 343–60.

Sharpe, F. R. and Lotka, A. J. (1911) A problem in age distribution. *Philosophical Magazine* **21**: 435–8.

Sharpe, P. J. H., Curry, G. L., DeMichele, D. W. and Cole, C. L. (1977) Distribution model of organism development times. *Journal of Theoretical Biology* **66**: 21–38.

Shoemaker, C. A., Smith, G. E. and Helgesen, R. G. (1986) Estimation of recruitment rates and survival from field census data with application to poikilotherm populations. *Agricultural Systems* **22**: 1–21.

Slade, N. A. (1977) Statistical detection of density dependence from a series of sequential censuses. *Ecology* **58**: 1094–102.

Slade, N. A. and Levenson, H. (1982) Estimating population growth rates from stochastic Leslie matrices. *Theoretical Population Biology* **22**: 299–308.

Smith, R. H. (1973) The analysis of intra-generation change in animal populations. *Journal of Animal Ecology* **42**: 611–22.

Southern, H. N. (1970) The natural control of a population of tawny owls (*Strix aluco*). *Journal of Zoology, London* **162**: 197–285.

Southwood, T. R. E. (1978) *Ecological Methods with Particular Reference to the Study of Insect Populations.* Chapman and Hall, London.

Southwood, T. R. E. and Jepson, W. F. (1962) Studies on the populations of *Oscinella frit* L. (Dipt.: Chloropidae) in the oat crop. *Journal of Animal Ecology* **31**: 481–95.

Stedinger, J. R., Shoemaker, C. A. and Tenga, R. F. (1985) A stochastic model of insect phenology for a population with spatially variable development rates. *Biometrics* **41**: 691–701.

Sykes, Z. M. (1969) Some stochastic versions of the matrix model for population growth. *Journal of the American Statistical Society* **64**: 111–30.

Taha, H. A. and Stephen, F. M. (1984) Modelling with imperfect data: a case study simulating a biological system. *Simulation* **42**: 109–15.

Tanner, J. M. (1962) *Growth at Adolescence.* Blackwell, Oxford.

Taylor, B. E. (1988) Analysing population dynamics of zooplankton. *Limnology and Oceanography* **33**: 1266–730.

Taylor, B. E. and Slatkin, M. (1981) Estimating birth and death rates of zooplankton. *Limnology and Oceanography* **26**: 143–58.

Tuljapurkar, S. D. (1982) Population dynamics in variable environments. 2. Correlated environments, sensitivity analysis and dynamics. *Theoretical Population Biology* **21**: 114–40.

Tuljapurkar, S. D. and Orzack, S. H. (1980) Population dynamics in variable environments. 1. Long run growth rates and extinction. *Theoretical Population Biology* **18**: 314–42.

Turner, H. N. and Young, S. S. Y. (1969) *Quantitative Genetics in Sheep Breeding.* Macmillan, Melbourne.

Usher, M. B. (1966) A matrix approach to the management of renewable resources, with special reference to selection forests. *Journal of Applied Ecology* **3**: 355–67.

Usher, M. B. (1969) A matrix model for forest management. *Biometrics* **25**: 309–15.

van Straalen, N. M. (1982) Demographic analysis of arthropod populations using a continuous stage-variable. *Journal of Animal Ecology* **51**: 769–83.

van Straalen, N. M. (1985) Comparative demography of forest floor Collembola populations. *Oikos* **45**: 253–65.

Vandermeer, J. (1978) Choosing category size in a stage projection matrix. *Oecologia* **32**: 79–84.

Varley, G. C. and Gradwell, G. R. (1960) Key factors in population studies. *Journal of Animal Ecology* **29**: 399–401.

Varley, G. C. and Gradwell, G. R. (1970) Recent advances in insect population dynamics. *Annual Review of Entomology* **15**: 1–24.

Varley, G. C., Gradwell, G. R. and Hassell, M. P. (1973) *Insect Population Ecology*. Blackwell, Oxford.

Wagner, T. L., Wu, H., Sharpe, P. J. H., Schoolfield, R. M. and Coulson, R. N. (1984a) Modelling insect development rates: a literature review and application of a biophysical model. *Annals of the Entomological Society of America* **77**: 208–25.

Wagner, T. L., Wu, H., Sharpe, P. J. H. and Coulson, R. N. (1984b) Modelling distributions of insect development time: a literature review and application of the Weibull function. *Annals of the Entomological Society of America* **77**: 475–87.

Wagner, T. L., Wu, H., Feldman, R. M., Sharpe, P. J. H. and Coulson, R. N. (1985) Multiple-cohort approach to simulating development of insect populations under variable temperatures. *Annals of the Entomological Society of America* **78**: 691–704.

Welch, S. M., Croft, B. A., Brunner, J. F. and Michels, M. F. (1978) PETE: an extension phenology modelling system for management of multi-species pest complex. *Environmental Entomology* **7**: 482–94.

Wilson, L. F. (1959) Branch 'tip' sampling for determining abundance of spruce budworm egg masses. *Journal of Economic Entomology* **52**: 618–21.

Woodward, I. O. (1982) Modelling population growth in stage-grouped organisms: a simple extension to the Leslie model. *Australian Journal of Ecology* **7**: 389–94.

Young, S. S. Y. and Wrensch, D. L. (1981) Relative influence of fitness components on total fitness of the two-spotted spider mite in different environments. *Environmental Entomology* **10**: 1–5.

Author index

Aitchison, J. 97
Aksnes, D. L. 49
Arnason, A. N. 32
Ashford, J. R. 43, 49, 64, 66

Bailey, V. A. 122
Baniuk, L. 32
Begon, F. J. 31
Begon, M. 45
Bellows, T. S. 49, 50, 79–83, 85, 86, 89, 95, 96, 99, 151
Benton, M. J. 45
Bernardelli, H. 101, 102, 103, 104, 108, 109
Berryman, A. A. 149, 150
Bhattacharya, C. G. 167
Birley, M. 44, 49, 50, 79–83, 85, 86, 89, 95, 96, 99
Blough, D. K. 162, 163
Blower, J. G. 31
Boyce, M. S. 109
Bradley, J. S. 2, 29
Braner, M. 26, 28, 41, 45, 51, 77
Brillinger, D. R. 153, 155, 156
Brown, K. M. 106
Brownie, C. 32
Burnham, K. P. 32, 40

Carter, N. 113
Caswell, H. 106, 108, 115
Cherrill, A. J. 45
Clobert, J. 32
Cochran, W. G. 15, 18, 19, 24
Connor, M. D. 163
Cooley, J. M. 27, 28, 49, 83
Cooper, R. J. 29
Cormack, R. M. 32
Croft, B. A. 2
Crosbie, S. F. 32
Crouse, D. T. 107, 108
Curry, G. L. 45

Dempster, J. P. 48

den Boer, P. J. 131, 143, 145, 146, 151
Dennis, B. 50, 97
Derr, J. A. 49
DeVries, P. G. 32
Dixon, W. J. 36
Dobson, A. J. 33
Dorazio, R. M. 115

Eberhardt, L. L. 32
Edmonson, W. T. 115, 117
Elkinson, J. S. 151

Fargo, W. S. 152
Ferris, H. 159, 161, 162
Fisher, R. A. 31
Ford, E. B. 31

Gabriel, W. 115
Gates, C. E. 32
Gradwell, G. R. 121, 122–3, 124, 138
Gurney, W. S. C. 152

Hairston, N. G. 26, 28, 41, 45, 49, 51, 77
Hart, R. C. 116
Hassell, M. P. 131, 143, 145, 150
Hayman, B. I. 126
Haynes, D. L. 48
Henderson, A. E. 126
Hiby, A. R. 108
Hoisaeter, T. 49
Holt, J. 166
Hughes, R. D. 115

Ito, Y. 131

Jackson, C. H. N. 31
Jepson, W. F. 48, 151
Jolly, G. M. 31, 32

Keen, R. 117
Kemp, W. P. 50, 51, 77
Kempton, R. A. 49, 64–79, 80, 83, 86
Kirchberg, M. 143, 145

Kiritani, K. 48, 52, 86, 90
Klomp, H. 5, 151
Kobayashi, S. 48
Kuno, E. 131

Lakhani, K. H. 43, 49
Latto, J. 143, 145
Lau, C. L. 67
Lefkovitch, L. P. 3, 5, 101, 103–5, 106, 110, 111, 113, 114
Leslie, P. H. 101, 102, 103, 104, 108, 109
Levenson, H. 109
Lewis, E. G. 101, 102, 103, 104, 108, 109
Lewis, W. M. 49
Liebhold, A. M. 151
Lih, M. P. 163, 166
Logan, J. A. 45
Longstaff, B. C. 108
Lotka, A. J. 101

Mackauer, M. 115
Manly, B. F. J. 26, 32, 34, 35, 36, 46, 47, 48, 49, 50, 51, 52, 62, 86, 90, 109, 110, 132, 137, 138, 149, 159
Mardia, K. V. 149
May, R. M. 152
McCullagh, P. 3, 36, 40, 97
McDonald, L. L. 26, 159
Mills, N. J. 50
Moloney, K. A. 105
Morris, R. F. 118, 122
Morrison, L. M. 29
Mullen, A. J. 108
Munholland, P. L. 27, 51, 67, 77

Nakasuji, F. 48, 52, 86, 90
Naranjo, S. E. 166
Nassar, R. 117
Nelder, J. A. 36, 40
Nicholson, A. J. 122, 149, 152
Nordheim, E. V. 109

Ord, K. 49
Orzak, S. H. 109
Osawa, A. 50, 51, 97
Otis, D. L. 31

Pajunen, V. I. 44
Paloheimo, J. E. 115, 116, 117
Plant, R. E. 103
Podoler, H. 124, 125, 126, 127, 138, 141, 148
Poethke, H. J. 143, 145

Pollard, J. H. 101, 103
Pontius, J. S. 94, 95, 96, 97, 99, 100

Qasrawi, H. 3, 4, 37, 42, 43, 44, 54, 55, 63, 69

Radford, P. J. 166, 167
Read, K. L. Q. 49, 64, 66
Reddingius, J. 131
Reed, T. E. 36
Richards, O. W. 46, 48
Rigler, F. H. 27, 28, 49, 83
Rogers, D. 126, 127, 138, 141, 148
Royama, T. 131, 137, 150
Ruesink, W. G. 49

Saunders, J. F. 49
Sawyer, A. J. 48, 166
Schaalje, G. B. 46
Schneider, S. M. 26, 159, 161, 162
Schull, W. J. 36
Seber, G. A. F. 31, 32
Seitz, A. 115
Service, M. W. 43, 49
Sharpe, F. R. 101
Sharpe, P. J. H. 45
Shoemaker, C. A. 50
Silvey, S. D. 97
Slade, N. A. 109, 131
Slatkin, M. 115
Smith, R. H. 126, 127, 138, 141, 148
Southern, H. N. 120
Southwood, T. R. E. 7, 31, 32, 48, 151
Stedinger, J. R. 50
Stephen, F. M. 163, 166
Sykes, Z. M. 109

Taha, H. A. 163
Tanner, J. M. 3
Taylor, B. E. 115
Tuljapurkar, S. D. 109
Turner, H. N. 126
Twombly, S. 49, 106

Usher, M. B. 105–7, 110, 114

van der Vaart, H. R. 46
Vandermeer, J. 105
van Straalen, N. M. 43, 44, 50, 71
Varley, G. C. 119, 121, 122, 123, 124, 131, 138, 143

Wagner, T. L. 45, 165

Waloff, N. 48
Way, M. J. 115
Welch, S. M. 2
Wilson, L. F. 7
Woodson, W. D. 152

Woodward, I. O. 108
Wrensch, D. L. 126

Young, S. S. Y. 126

Subject index

Aedes cantans 43
Aeneolamia varia saccharina 44
Aleurotrachelus jelinekii 151
Anasta tristis 152

Beta distribution, *see* Distribution
Binomial distribution, *see* Distribution
BMDP 36
Breast development 3, 5, 91, 98–9
Brown planthopper, *see Nilaparvata
    lugens*
Bupalus piniarius 5–6, 151

Callosobruchus chinensis 86–90, 92–3,
    95–6
Caretta caretta 106–8
Ceratitis capitata 103
Chi-squared distribution, *see*
    Distribution
Choristoneura fumiferana 8
Chorthippus
    brunneus 29
    parallelus 3, 4, 29, 42–3, 55–62, 63–4,
    69–70, 81–2
Cigarette beetle, *see Lasioderma
    serricorne*
Cluster sampling, *see* Sampling
Confidence interval
    for population mean 10
    for population proportion 11
    for population total 10
Continuous recruitment 3, 101–17
Corn rootworm, *see Diabrotica barberi*
Corythucha pallipes 27
Cratichneumon culex 119
Cucurbita pepo 152
Cyzensis albicans 119

Daphnia publicaria 115
Degree days, *see* Physiological time
Degrees of freedom 37–8
Delayed density-dependent mortality
    121, 131–2, 146, 148–50

Dendroctonus frontalis 163–6
Density-dependent survival 121–2, 128–
    32, 152
Development variable 50, 51, 77
Deviance 37–41, 70, 71, 79
    see also Heterogeneity factor
Diabrotica barberi 166
Diffusion process 77
Distribution
    beta 109
    binomial 95
    chi-squared 37, 40
    Erlangian 64–5, 159
    F 40
    gamma 26, 51, 65–9, 71, 73, 78–9, 85,
    109
    inverse normal 65, 73, 77, 78, 85
    logistic 50
    lognormal 65, 109
    multinomial 34–6, 67
    normal 45, 50, 51, 65, 73, 77, 78, 85,
    108, 117, 167
    Poisson 3, 25–7, 33–6, 39–41, 47, 51,
    56–7, 62, 64, 67, 80, 82, 85, 117
    product multinomial 34, 36
    Weibull 65, 80–1, 89

Egg to adult ratio 3, 115–17
Erlangian distribution, *see* Distribution
Estimation
    bias in estimates 58, 64
    combined ratio estimation 19
    of insect damage 11–12
    of means 10, 14, 18
    of mean time of entry to stage 1 65, 70,
    71, 78
    of numbers entering stages 54, 56, 63,
    64, 70, 71, 78, 81, 88
    of population total 10, 14, 18
    of proportions 10–11, 14–15
    of size of insect population 15–17, 19–
    21, 31–2

of stage durations 45–6, 54, 56, 63, 64,
    70, 71, 78, 82, 88, 90, 91, 94–6,
    97–9, 162
of stage-specific survival rates 53, 56,
    58, 63, 68, 88, 90, 162
of survival parameter 53, 56, 58, 63,
    68, 70, 71, 78, 82
ratio 18–19
regression 22–3
separate ratio estimation 19
Examples
    analysis of a generation of *Orchesella
        cincta* 70–3, 78–9, 85
    analysis of a single cohort of
        Callosobruchus chinensis 85–90,
        95–6
    artificial data 99–100, 116–17
    breast development of New Zealand
        schoolgirls 97, 98–9
    Chorthippus parallelus on East
        Budleigh Common 55–62, 63–4,
        69–70, 81–2
    counting fir trees in a forest 11–12
    grey pup seal 166–7
    key factor analysis
        on the pine looper 151
        on the tawny owl 124–5, 127–8, 138,
            146–8
        on the winter moth 123–5, 127, 137,
            140–6, 150
    laboratory population of the cigarette
        beetle 110–14
    modelling population dynamics of
        loggerhead turtles 106–8
    nematode 159–62
    pink bollworm moth 162–3
    sampling
        arthropods in a forest 29–30
        four grasshoppers on a common site
            29
        trees in a forest 19–21, 22–3
        zooplankton in Teapot Lake 27–8,
            83–5
    sheep blowfly 152–9
    southern pine beetle 163–6
    stratified sampling of an insect
        population 15–17
Extraneous variance, *see* Heterogeneity
    factor

Finite population correction 10
Froghopper, *see Aeneolamia varia
    saccharina*

Gamma distribution, *see* Distribution
GLIM 32, 36
Goodness of fit 25–6, 37–41, 47, 61–2
Grey pup seal, *see Halichoerus grypus*
Gypsy moth, *see Lymantria dispar*

Halichoerus grypus, 166–7
Heterogeneity factor 3, 25–6, 39–41, 62,
    71, 82, 85, 89–90, 97

Insecticide spraying 44, 80, 162
Inverse normal distribution, *see*
    Distribution

Jolly–Seber model 31–2

Kalman filter 155, 162
Key factor analysis 5–6, 118–51
Key factor data
    on the pine looper 6
    on the tawny owl 120
    on the winter moth 119

Lasioderma serricorne 5, 101, 110–14
Leslie matrix 102–3
Likelihood function 33–7
Line intercept sampling 32
Loggerhead turtles 106–8
Logistic function 97, 100
Log–log function 97, 98, 100
Lognormal distribution, *see* Distribution
Lucilia cuprina 152–9
Lymantria dispar 151

Mark-recapture, *see* Sampling
Maximum likelihood 2, 25–6, 31, 33–7,
    40, 49, 50, 51, 64–5, 69, 83, 155, 162
MAXLIK computer program 36, 67, 70,
    71, 80, 82, 89–90, 161–2
Mediterranean fruit fly, *see Ceratitis
    capitata*
Migration 43
Multi-cohort data 3, 42–85, 86
Multinomial distribution, *see*
    Distribution
Multi-stage sampling, *see* Sampling
Myzus persicae 115

Nematodes, *see Paratrichodorus minor*
Nilaparvata lugens 166
Normal distribution, *see* Distribution
Null model 139, 140

Omocestus
  lineatus 29
  viridulus 29
Operophtera brumata 118, 119, 121,
    123–4, 128–9, 131, 137–8, 140–6,
    150
Orchesella cincta 43–4, 70–3, 78–9

Panaxia dominula 31
Parasitism 118, 119, 122, 139, 149, 151
Paratrichodorus minor 159–62
Pectinophora gossypiella 162–3
PETE 2
Phase portraits 150
Phenology 2
Physa ampullacea 106
Physiological time 44–5, 50, 51, 69, 71,
    85, 115, 159
Pine looper, *see Bupalus piniarius*
Pink cotton bollworm, *see Pectinophora
    gossypiella*
Plistophora operopterae 119
Poisson distribution, *see* Distribution
Population
  reproducing 3, 101–17
  sampled 7, 30
  target 7, 30
  variance 10
Predation 139, 146
Principal component analysis 138
Product multinomial, *see* Distribution

Random number table 7–8
Rate summation method 165
Ratio estimation, *see* Estimation
Regression
  linear 22, 48, 50, 104–5, 106, 126, 127,
    128, 131, 133, 156, 157
  non-linear 49, 50, 80
Removal sampling, *see* Sampling
Rice weevil, *see Sitophilus oryzae*

Sample size determination 12–13, 15
Sampled population 7, 30
Sampling
  arthropods 29–30
  biased 26, 159
  cluster 23, 30
  encounter 30
  errors with key factor data 140, 146
  fraction, *see* Finite population
    correction

frame 8
grasshoppers 29
line-intercept 32
line-transect 30, 32
mark-recapture 29, 30, 31–2
multi-stage 24
removal 30–1
simple random sampling 8, 26
species assemblages 29
stage-frequency data 24–7
stratified 13–15, 25, 30
strip-transect 32
systematic 23–4
transect 32
unit 8, 30
with and without replacement 8
zooplankter 27–8
SAS (Statistical Analysis System) 32, 36
Sheep blowfly, *see Lucilia cuprina*
Simple random sampling, *see* Sampling
Simulation 46–7, 51–2, 56–62, 64, 83–5,
    99–100, 104, 110, 112, 114, 115,
    139–48, 149–50, 151, 152
Single cohort data 3, 42, 86–100
Sitophilus oryzae 108
Size classes 43, 44
Skistodiaptomus oregonensis 27–8, 40–1,
    83–5
SLAM (simulation language for
    alternative modelling) 152
Southern pine beetle, *see Dendroctonus
    frontalis*
Spatial information on a population 151,
    163
SPBMODEL 163–6
Spider mites 126
Spiral plot 131–3, 146, 148–50
Spruce budworm *see Choristoneura
    fumiferana*
Squash bug, *see Anasta tristis*
Squash plant, *see Cucurbita pepo*
Stage durations, *see* Estimation
Stage-frequency data
  defined 2
  on breast development 5
  on *Callosobruchus chinensis* 87
  on *Chorthippus parallelus* 4
  on *Lasioderma serricorne* 5, 111
  on *Orchesella cincta* 44
Stratified sampling, *see* Sampling
Strip-transect sampling 32
Strix aluco 118, 120, 121, 124, 128, 130–1
Systematic sampling, *see* Sampling